D0181772

ALSO BY DAVID QUAMMEN

NONFICTION

*The Song of the Dodo*

*The Flight of the Iguana*

*Natural Acts*

FICTION

*Blood Line*

*The Soul of Viktor Tronko*

*The Zolta Configuration*

*To Walk the Line*

# WILD THOUGHTS *from* WILD PLACES

David Quammen

A TOUCHSTONE BOOK
PUBLISHED BY SIMON & SCHUSTER

TOUCHSTONE
*Rockefeller Center*
*1230 Avenue of the Americas*
*New York, NY 10020*



*First Touchstone Edition 1999*

TOUCHSTONE *and colophon are registered trademarks of Simon & Schuster Inc.*
*Designed by Brooke Zimmer*
*Manufactured in the United States of America*

*1 3 5 7 9 10 8 6 4 2*

*The Library of Congress has cataloged the Scribner edition as follows:*
*Quammen, David, date.*
*Wild thoughts from wild places / David Quammen.*
*p. cm.*
*Includes bibliographical references.*
*1. Natural history. 2. Quammen, David—Journeys.*
*I. Title.*
*QH81.Q18 1998*
*508-dc21 97-29090*
*CIP*
*ISBN 0-684-83509-6*
*ISBN 0-684-85208-X (Pbk)*

To Kay and Sallie

# Contents

## III. THE MOUNTAINS



## IV. THE HEART

# WILD THOUGHTS *from* WILD PLACES

# Introduction

The title of this book alludes to a scarce resource. Wild places, in the ordinary sense of that phrase, are in preciously short supply on planet Earth at the end of the twentieth century.

The great zones of forest and swamp have been reduced to small, tattered remnants of what they once were. The fiercest deserts have been split by roads and planted with artificial oases. The widest seas have been effectively narrowed, the highest mountains effectively lowered, the coldest polar regions made effectively warmer and more hospitable, not simply by adventurous human visitors but by the cultural and technological trappings that modern humans bring. For instance, nowadays you might step out of a dugout canoe at the Amazon headwaters and meet an Indian man wearing a red feather through his nose and a gimme cap reading OKLAHOMA SOONERS. You might gaze into the sky over the highlands of New Guinea and see a helicopter, bringing in two oil geologists along with their wall tent, their survey equipment, their Macintosh, and their food, including a two-month supply of Pringles. You might enter a little village café on the island of Kosrae, in the southwestern Pacific, and find several local children watching the Three Stooges on a VCR. To a considerable degree, these and other wild places have been tamed.

Some people celebrate that fact—Hooray, humanity conquers the wilderness, making life safer and more comfortable for everybody. Other people, including me, regret it. I suppose you could say: We *have the luxury* of regretting it. But that's a complicated argument, and not one I want to pursue here. My purpose at the moment is more modest: to welcome you to this book of ruminations, and to intimate a definition of "wild places" that's a little more stretchy and inclusive.

Wildness, in this broader sense, inheres in any geographical or emotional context that remains unpolluted by absolute safety and certainty. Take the foothills of Burbank, with their coyote population—as described in the essay here titled "To Live and Die in L.A." Or the peculiar canyon ecology of midtown Manhattan, fostering evolutionary adaptations in the postmodern pigeon—as described in "Superdove on 46th Street." Or the death of a beloved curmudgeon named Edward Abbey—as described in "Bagpipes for Ed." I can report finding a certain richly threatening wildness in each of those contexts, those geographical and emotional places, which I've tried to share.

My own discovery of wildness, in the most discomfitingly personal sense, dates back about twenty-five years to an episode in the mountains of north-central Wyoming. The memory of that episode is indelible to me, though the story isn't especially dramatic. I had driven out west just after graduate school, in quest of purgation and a new life and trout. I was alone in my Volkswagen bus, carrying almost everything I owned, which consisted mostly of paperback books, a skillet, a pair of waders, and a fly rod, the last of which I was just learning to use. My life stretched before me as a mysterious, undifferentiated vista, like the plains of Nebraska at sunset, vacant of plan or intention except that I knew I wanted to write novels. I was reading *War and Peace,* in a two-volume Penguin edition, and my submersion in that book reflects the fact that for the first time since boyhood I had an abundance of free time. I was nearly broke, but my prospects and my liberty seemed boundless.

I camped for a week in the Bighorn Mountains, above Sheridan, living out of the bus and fishing a fork of the Tongue River that meandered past my campground. Since the stream was tiny and I'd been sharing it with other fishermen, one day I decided to hike downstream a few miles to where it converged with another fork, far from any campground or road. (Consulting a map now, half a lifetime later, I see that those tributaries must have been the North Tongue and Fool's Creek, the second of which seems apropos, in light of my behavior.) I jammed my waders into a day pack along with a canteen, matches, and a box of raisins, and carried my fly rod in my hand. Within a mile the trail evaporated and I found myself bushwhacking along the stream. Within two miles the route had become clotted with thick timber, underbrush, and deadfalls, interrupted occasionally by barricades of scree. The going had gotten slow, but I plodded on.

In midafternoon I passed an old derelict cabin and made a mental note of it. In late afternoon, fighting through brush, I reached the little confluence—which turned out, disappointingly, despite the doubled volume of water, to offer no particularly good fishing spots. Trying a Black Gnat pattern and a few other flies, I caught nothing. The sun was now low and I was hungry. So I impaled a live grasshopper on a hook, dangled it into a pool too obstructed for me to hit with a cast, and landed a single small trout. Then I started walking out. Dusk caught me in the vicinity of the derelict cabin. By this time it was clear that I'd better stay here for the night. I knew that busting the thickets and crossing the scree piles in darkness would be ugly work, and foolhardy; one bad step, one wet rock, and I could break a leg. And I realized, rather belatedly, that a broken leg under these circumstances could be fatal, since no one in the world knew within five hundred miles or three weeks where I might be. I had made no local friends; I had designed no fallback scenario. I was only a half dozen miles from a road, but I had never been more alone. Although I'd been too stupid to think of that earlier, now I could think of nothing else. I had put myself into a context that, narrow as it was, trivial as its dangers might be, seemed as wild as any I've ever known.

I built a fire. I cooked the trout on a stick, and it tasted delicious. The raisins were good too. I lay down on a plank bunk inside the cabin until the pack rats, crawling over my legs, spooked me. Then I went back outside and tried to sleep in my waders, thinking they would insulate me like a sleeping bag, but they were too wet and clammy. So I rebuilt the fire and curled up near it, scorching the elbow out of my dirty wool sweater. While I hunkered and shivered, the pack rats chewed sections from my fly line and hauled them away. In the morning, the glorious morning, I hiked out, placing each footstep with extraordinary care. Safely back at the trailhead, I was so giddy with relief that I let my fly rod fall in front of the Volkswagen bus and then drove over it.

This was my little baptism into a new life, a new set of perspectives, in the Wild West. The whole conversion experience is mentioned passingly in "Synecdoche and the Trout," the first piece in this collection, and again in "Strawberries Under Ice," the last. It's no coincidence that those two essays, concerning my own most beloved wild place, bracket the rest.

Between them lies a sampler of other travels, misadventures, small insights, research efforts, exercises in scientific explication and unscientific noodling, portraits of people who have fascinated and impressed me, sporting narratives, eulogies, reminiscences, valentines, warnings, second thoughts, and willful provocations, all of which have been previously published, mostly in magazines. The order in which they're presented here—within four loose wild-place rubrics, the River, the City, the Mountains, the Heart—is not chronological, nor necessarily the order in which you should read them. Although there's a certain logical progression among some of the essays as arranged (familiarizing you with whitewater terminology from one essay to another within the River section, for instance, or with the peculiar character of telemark skiing from one essay to another within the Mountains), I invite you to jump around, dipping into the various sections as your tastes or your mood might dictate. The real point of even those river-washed and mountain-hung stories, after all, is not to convey athletic information but to offer some observations and thoughts about interesting people in intense situations.

The majority of these pieces appeared originally as installments of the column I wrote for fifteen years, under the title "Natural Acts," for *Outside* magazine. (One of them, "Superdove on 46th Street," is my final "Natural Acts" column, which ran in the March 1996 issue.) Most of the rest were done as feature assignments, either for *Outside* or for other magazines, such as

*Rolling Stone* or *Powder.* I list all the details of first publication in "Notes and Provenance," at the back. Before leaving this subject, though, I want to say one other thing about magazines.

Magazines, and especially the "slick" magazines that safely tenured academics and high-minded literati sometimes scorn, offer a vast richness of opportunity to imaginative nonfiction writers—and therefore also to imaginative nonfiction readers. These magazines get a bad rap. The sharp criticisms and easy condescension leveled against them are generally unconsidered, supercilious, blindered to the tricky relations between writer and audience, also to those between writer and creditors, and no more than about half accurate. Because they print on glossy paper, carry advertisements for fancy sunglasses and Rollerblades and miracle-fiber underwear, and devote a portion of their pages to items and stories that might politely be called fatuous junk, magazines such as *Outside* are too often dismissed as intellectually or literarily negligible. But stop and think about it, and you'll remember that the *New York Times* also devotes a sizable portion of its pages to fatuous junk. My point is that magazines, like newspapers, should be judged by the best and most substantial of what they offer, not by the worst and most trivial stuff that serves toward helping them meet the payroll. The world is big and wondrous, full of odd sights and strange beasts and weird noises and charmingly demented people with great stories to tell and (despite the increasing scarcity) more than a few wild places—wild places to which the sainted editors of *Outside* and certain other slick magazines are forever sending writers, often with little more instruction than to observe carefully, think hard, and then put something honest on paper. Furthermore, these magazines reward their writers with decent, life-sustaining payment. (I'm not being metaphorical here; I mean money.) This is not negligible. This is a service that delivers meat—thinly sliced meat that's there for whoever cares to find it, like kosher pastrami hidden in a Wonder bread sandwich.

One of those lucky, trusted writers has been me. If it wasn't so, I would have missed the chance to risk drowning myself on the Futaleufu River—as described in "Grabbing the Loop"—or to find meaning over a plate of stir-fried mountain lion—as described in "Eat of This Flesh." I would have missed seeing

Hell's Hole on the Ocoee River during a whitewater world championship, missed hearing currawongs caw on a high moor in northwestern Tasmania, missed getting snowbound with a crew of ICBM launch officers, and missed the avalanche triggered by Karl Birkeland near Lionhead Peak. I would have missed aspects of the world that libraries and telephones and highbrow journals could never show me. I would have missed the chance of offering you this collection of far-flung reports and reckless notions. I might have had to stay home, God forbid, and look for a steady job.

[1]

# *The* RIVER

# *Synecdoche and the Trout*

It's a simple question with a seemingly simple answer: "Why do you live in Montana?"

Repeatedly over a span of some years you have heard this, asked most often by people who know you just well enough to be aware of the city where you grew up, the tony universities you attended, and a few other bits of biographical detail on the basis of which they harbor a notion that you should have taken your place in New York café society or, at least, an ivy-adorned department of English. They suspect you, these friends do, of hiding out. Maybe in a way they are right. But they have no clear sense of what you are hiding from, or why, let alone where. Hence their question.

"The trout," you answer, and they gape back blankly.

"The trout," they say after a moment. "That's a fish."

"Correct."

"Like lox."

"In some ways similar."

"You like to go fishing. *That's* why you live out there? *That's* why you spend your life in a place without decent restaurants or bookstores or symphony orchestras, a place halfway between Death Valley and the North Pole? A place where there's no espresso, and the *Times* comes in three days late by pontoon plane? Do I have this straight, now? It's because you like to go *fishing?*"

"No," you say. "Only partly. At the beginning, that was it, yes. But I've stayed all these years. No plans to leave."

"You *went* for the fishing, but you *stayed* for something else. Aha."

"Yes. The trout," you say.

"This is confusing."

"A person can get too much trout fishing. Then it cloys, becomes taken for granted. Meaningless."

"Again like lox."

"I don't seem to fish nearly as much as I used to."

"But you keep talking about the trout. You went, you stayed, the trout is your reason."

"The trout is a synecdoche," you say, because these friends are tough and verbal and they can take it.

A BIOLOGIST would use the term *indicator species.* Because I have the biases of a literary journalist, working that great gray zone between newspaper reporting and fiction, engaged every day in trying to make facts not just talk but yodel, I speak instead of synecdoche. We both mean that a trout represents more than itself—but that, importantly, it does also represent itself.

"A poem should not mean/But be," wrote Archibald MacLeish, knowing undeniably in his heart that a good poem quite often does both. Likewise a trout.

The presence of trout in a body of water is a discrete ecological fact that nevertheless signifies certain things.

It signifies a particular complex of biotic and chemical and physical factors, a standard of richness and purity, without which that troutly presence is impossible. It signifies aquatic nutrients like calcium, potassium, nitrate, phosphate; signifies enough carbon dioxide to nourish meadows of algae and to keep calcium in solution as calcium bicarbonate; signifies a prolific invertebrate fauna (Plecoptera, Trichoptera, Diptera, Ephemeroptera), and a temperature regime confined within certain daily and annual extremes. It also signifies clear pools emptying down staircases of rounded boulders and dappled with patterns of late-afternoon shade cast by chrome yellow cottonwood leaves in September. It signifies solitude so sweet and pure as to bring an ache to the sinuses, a buzz to the ears. Loneliness and anomie of the most wholesome sort. It signifies dissolved oxygen to at least four or five parts per million. It signifies a good possibility of osprey, dippers, and kingfishers, otters and water shrews, heron; and it signifies *Oncorhynchus clarki, Oncorhynchus mykiss, Salmo trutta.* Like a well-chosen phrase in any poem, MacLeish's included, the very presence of trout signifies at more than one level. Magically, these creatures are literal and real. They live in imagination, memory, and cold water.

For instance: I can remember the first trout I ever caught as an adult (which was also the first I ever caught on a fly), and precisely what the poor little fish represented to me at that moment. It represented (a) dinner and (b) a new beginning, with a new sense of self, in a new place. The matter of dinner was important, since I was a genuinely hungry young man living out of my road-weary Volkswagen bus with a meager supply of groceries. But the matter of selfhood and place, the matter of reinventing identity, was paramount. My hands trembled wildly as I took that fish off the hook. A rainbow, all of seven or eight inches long. Caught on a Black Gnat pattern, size 12, tied cheaply of poor materials somewhere in the Orient and picked up by me at Herter's when I had passed through South Dakota. I killed the little trout before it could slip through my fingers and, heartbreakingly, disappear. This episode was for me equivalent to the one in Faulkner's "Delta Autumn," where blood from a fresh-killed buck is smeared on the face of the boy. *I slew you,* the boy thinks. *My bearing must not shame your quitting life,* he understands. *My conduct for ever onward must become your death.* In my own case, of course, there was no ancient Indian named Sam Fathers serving as mentor and baptist. I was alone and an autodidact. The blood of the little trout did not steam away its heat of life into the cold air, and I smeared none on my face. Nevertheless.

The fish came out of a creek in the Bighorn Mountains of north-central Wyoming, and I was on my way to Montana, though at that moment I didn't yet know it.

Montana was the one place on Earth, as I thought of it, farthest in miles and spirit from Oxford University, yet where you could still get by with the English language, and the sun didn't disappear below the horizon for days in a row during midwinter, and the prevailing notion of a fish dinner was not lutefisk. I had literally never set foot within the boundaries of the state. I had no friends there, no friends of friends, no contacts of any sort, which was fine. I looked at a map and saw jagged blue lines, denoting mountain rivers. All I knew was that, in Montana, there would be more trout.

Trout were the indicator species for a place and a life I was seeking.

I went. Six years later, rather to my surprise, I was a professional fishing guide under license from the Montana Department of Fish, Wildlife, and Parks. My job was to smear blood on other young faces. *I slew you. My bearing must not shame your quitting life.* Sometimes it was actually like that, though quite often it was not.

*Item.* You are at the oars of a fourteen-foot Avon raft, pushing across a slow pool on the Big Hole River in western Montana. An August afternoon. Seated in front of you is an orthopedic surgeon from San Francisco, a pleasant man who can talk intelligently about the career of Gifford Pinchot or the novels of Evelyn Waugh, who is said to play a formidable game of squash, and who spends one week each year fishing for trout. In his right hand is a Payne bamboo fly rod that is worth more than the car you drive, and attached to the rod is a Hardy Perfect reel. At the end of the doctor's line is a kinked and truncated leader, and at the end of the leader is a dry fly that can no longer by even the most technical definition be considered "dry," having been slapped back and forth upon and dragged through several miles of river. With this match of equipment to finesse, the good doctor might as well be hauling manure in the backseat of a Mercedes. Seated behind you is the doctor's wife, who picked up a fly rod for the first time in her life two hours earlier. Her line culminates in a fly that is more dangerous to you than to any fish in Montana. As you have rowed quietly across the glassy pool, she has attacked the water's surface like a French chef dicing celery. Now your raft has approached the brink of a riffle. On the Big Hole River during this late month of the season, virtually all of the catchable trout cluster (by daylight, at least) where they can find cover and oxygen—in those two wedges of deep still water flanking the fast current at the bottom of each riffle. You have told the doctor and his wife about the wedges. There, *those,* you have said. Cast just across the eddy line, you have said. Throw a little slack. We've got to hit the spots to catch any fish, you have said in the tactfully editorial first-person plural.

As your raft slides into this riffle, the doctor and his wife become tense with anticipation. The wife snags her fly in the rail rope along the rowing frame, and asks sweetly if you would free it, which you do, grabbing the oars again quickly to avoid hitting a boulder. You begin working to slalom the boat through the riffle. The wife whips her fly twice through the air before sinking it into the back of your straw cowboy hat. She apologizes fervently. Meanwhile, she lets her line loop around your right oar. You take a stroke with the left oar to swing clear of a drowned log, then you point your finger over the doctor's shoulder: "Remember, now. The wedges." He nods eagerly. The raft is about to broadside another boulder, so you pull hard on both oars and with that motion your hat is jerked into the river. The doctor makes five false casts, intent on the wedges, and then fires his line forward into the tip of his own rod like a handful of spaghetti hitting a kitchen wall. He moans. The raft drops neatly out of the riffle, between the wedges, and back into dead water.

*Item.* You are two days along on a wilderness float through the Smith River canyon, fifty miles and another three river-days from the nearest hospital, with cliffs of shale towering hundreds of feet on each side of the river to seal you in. The tents are grouped on a cottonwood flat. It's dinner hour, and you have just finished a frigid bath in the shallows. As you open your first beer, a soft-spoken Denver architect walks back into camp with a size 14 Royal Wulff stuck past the barb into his lower eyelid. He has stepped behind another fisherman at precisely the wrong moment. Everyone looks queasily at everyone else, but the outfitter—who is your boss, who is holding his second martini, and whose own nerves are already frazzled from serving as chief babysitter to eight tourist fishermen—looks pleadingly at you.

With tools from your fishing vest (a small pair of scissors, a forceps, a loop of leader) you extract the fly. Then you douse the architect's wound with what little remains of the outfitter's gin.

*Item.* Three days down the Smith on a different trip, under a cloudless July sky, you are drifting, basking comfortably in the heat, resting your oars. In your left hand is a cold Pabst Blue Ribbon. In place of your usual T-shirt, you are wearing a new yellow number that announces with some justice, "Happiness Is a Cold Pabst." On your head, in place of the cowboy straw, is a floppy cloth porkpie in a print of Pabst labels. In the bow seat of your raft, casting contentedly to a few rising trout, is a man named Augie Pabst, scion of the family. Augie, contrary to all your expectations, is a sensitive and polite man, a likable fellow. Stowed in your cargo box and your cooler are fourteen cases of Pabst Blue Ribbon, courtesy. You take a deep gulp of beer, you touch an oar. Ah yes, you think. Life in the wilderness.

*Item.* You are floating a petroleum engineer and his teenage son through the final twelve miles of the Smith canyon, which is drowsy, meandering water not hospitable to rainbow trout but good for an occasional large brown. The temperature is ninety-five, the midday glare is fierce, you have spent six days with these people, and you are eager to be rid of them. Three more hours to the take-out, you tell yourself. A bit later you think, Two more hours. The petroleum engineer has been treated routinely with ridicule by his son, and evidently has troubles also with his wife. The wife is along on this trip but she doesn't fish; she doesn't seem to talk much to her husband; she has ridden a supply boat with the outfitter and spent much of her time humming quietly. You wonder if the petroleum engineer has heard of Hemingway's Francis Macomber. You are sure that the outfitter hasn't and you suspect that the wife has. The engineer says that he and his son would like to catch one large brown trout before the trip ends, so you tell them to tie on Marabou Muddlers and drag those billowy monstrosities through certain troughs. Fifteen minutes later, the boy catches a large brown. This fish is eighteen inches long and broad of shoulder—a noble and beautiful animal that the Smith River has taken five years to grow. The father tells you to kill it—"Yeah, I guess kill it"—they will want to eat it, just this one, at the hotel. Suddenly you despise your job. You despise this man, but he is paying your wage and so he has certain prerogatives. You kill the fish, pushing your thumb into its mouth and breaking back the neck. Its old sharp teeth cut your hand.

The boy is a bad winner, a snot, taunting his father now as the three of you float on down the river. Half an hour later, the father catches a large brown, this one also around eighteen inches. You are pleased for him, and glad for the fish, since you assume that it will go free. But the father has things to prove to the wife as well as to the son, and for the former your eyewitness testimony of a great battle, a great victory, and a great act of mercy will not suffice. "Better keep this one too," he says, "and we'll have a pair." You detest this particular euphemistic use of the word *keep.* You argue tactfully but he pretends not to hear. Your feelings for these trout are what originally brought you out onto the Smith River and are what compel you to bear the company of folk like the man and his son. *My conduct for ever onward must become your death.* The five-year-old brown trout is lambent, spotted with orange, lithe as an ocelot, swirling gorgeously under water in your gentle grip. You kill it.

I DON'T guide anymore. I haven't renewed my license in years. My early and ingenuous ideas about the role of a fishing guide turned out to be totally wrong: I had imagined it as a life rich with independence, and with a rustic sort of dignity, wherein a fellow would stand closer to these animals he admired inordinately. I hadn't foreseen that it would demand the humility of a chauffeur and the complaisance of a pimp.

And I don't seem to fish nearly as much as I used to. I have a dilemma these days: I dislike killing trout but I believe that, in order to fish responsibly, to fish conscionably, the fisherman should at least occasionally kill. Otherwise he can too easily delude himself that fly fishing is merely a game, a dance of love, played in mutual volition and mutual empathy by the fisherman and the trout. Small flies with the barbs flattened are an excellent means for allowing the fisherman's own sensibilities to be released unharmed—but the fish themselves aren't always so lucky. They get eye-hooked, they bleed, they suffer trauma and dislocated maxillae and infection. Unavoidably, some die. For them, it is not a game, and certainly not a dance. On some days I feel it's hypocritical to profess love for these creatures while endangering and abusing them so wantonly; better to enjoy the thrill of the sport honestly, kill what I catch, and stop fishing when I've had a surfeit of killing. On other days I do dearly enjoy holding them in the water, gentling them as they regain breath and balance and command of their muscles, then watching them swim away. The dilemma remains unresolved.

"Yet each man kills the thing he loves," wrote Oscar Wilde, and I keep wondering how a person of Wilde's urban and cerebral predilections knew so goddamn much about trout fishing.

"Why do you live in Montana?" people ask. For the trout, I answer. "Oh, you're one of those fanatical fisherman types?" No, not so much anymore, I say. It's just a matter of knowing that they're here.

# *Time and Tide on the Ocoee River*

Chris Spelius is inventing a boat. It's a whitewater kayak, only different. At the moment, in mid-July, it exists nowhere but in his mind's eye. Because the World Whitewater Rodeo Championships will occur in October, just three months away, time is short. The task that he faces is formidable: Think up this new boat, persuade the Dagger Canoe Company to design it and build it, then use it himself to win the men's kayak championship. Fat chance at his age, forty-one. Still, Spelius loves a severe challenge. He inhales the pheromones of competition the way others inhale barbecue smoke. He's a big guy, broad-shouldered and blond, with a wide friendly grin and a square Pete Rose jaw, and his enthusiasm radiates outward like heat off a brick oven. As we sit in a small-town café near the bank of a famous river in Idaho, over the aftermath of greasy rib steaks, he describes the logic behind this hypothetical craft. He wants me to comprehend its arcane merits, to share the thrill of a technological breakthrough. He begins folding his paper place mat into the shape of an avant-garde kayak.

The nose will be flat, he explains. Like so. The tail, also radically flat. Like so. He folds up another place mat and sets it before him on the table. Okay now this one, he says earnestly, is the hole.

A hole, in whitewater parlance, is a recirculating maw of foam. Fast water pours over a sharp drop, tumbles back on itself, gurgles down, bubbles up, and tumbles back on itself again, creating a grabby trough in the river's surface. It swallows and reswallows its own water, like a vortex laid horizontal, and sucks floating objects toward it from both above and below, takes hold of them, keeps them. Its gape shows like a frothy smirk or, depending on viewpoint, a frown. Cautious paddlers make a point of going around. Others make a point of going in. There's a certain chaotic equilibrium that can be found, toyed with, enjoyed. The sensation of riding a hole is roughly akin to sitting the back of a surly bull driven wacko by a flank strap, and so kayakers have borrowed the word *rodeo.*

Approaching the notional hole from downstream, Spelius drives the nose of his paper kayak straight into it. At contact, he tilts the boat upright. This represents an *ender,* one of the basic maneuvers of whitewater rodeo. The sinking water takes the bow down, and the boat stands on end. It's only the start of what he means to demonstrate.

Drop an inexperienced kayaker into a good sticky hole, and you have an emergency. Drop in a fellow like Spelius, and you have a performance. Add a loudspeaker and judges and one or two newspaper photographers and an audience—preferably an audience that includes "babes," Spelius insists, in his teasingly retrograde language—and you have a whitewater rodeo. Music will blare, adrenaline will flow, talented athletes will take their turns in the maw. They'll do enders and cartwheels and pirouettes, they'll twirl their paddles or toss them away, they'll execute 360-degree spins and barrel rolls barehanded, impossible tricks made to look easy. They'll dance with the water, following its lead, slipping its grip, diving and emerging, spitting out mouthfuls of river, cavorting like porpoises at Sea World. No one will drown, probably. Prizes will be awarded—simple trophies, no cash—and everyone will go home with stuffed sinuses. This is the sport that Chris Spelius, over the past fifteen years, has helped to create.

His paper boat rises vertical over the hole. There's wide exposure here on the deck, he explains, so that the nose will go down like a shovel. You start your ender, you lift your right knee, you move into your pirouette, he tells me, as though that much were perfectly easy.

Now here's the sweet part, he says, eyes bright, as he twists the boat sideways. The flat nose, in silhouette, translates to a narrow blade. Meaning: not much water resistance as you pirouette over your ender. So the subsurface current doesn't push the boat out of the hole. One ender, then a pirouette, then with luck and a reaching reverse stroke you can drop straight back into the hole. Another ender, another pirouette, dropping again into the hole; and then a cartwheel, a barrel roll, another ender, another pirouette, a full spin, another cartwheel, whatever—all linked together. An entire sequence of *retendo* moves, he says, bringing me in on his favored bit of jargon, a portmanteau version of *ender* and *retainment.* Do you get it? he asks. Retendos are the future of whitewater rodeo. And the future is here.

I get it, but only on hearsay. This is the province of an elite few. My own visits to that zone of chaotic equilibrium are clumsy though exhilarating, and they don't involve cartwheels—not intentional ones, anyway. The sort of fandango that Spelius dreams of is far beyond my scope, and it coincides with a cautious kayaker's nightmare: to paddle into a hole and stay there forever, tumbling ass over teakettle.

"I'm not gonna jump on a trampoline and do somersaults," he says. "I'm not gonna do flips on skis. I'm not gonna jump out of an airplane and do flips. But in a kayak . . ." His voice trails away wistfully, then returns with a firm bit of self-knowledge. "Water," he says, "is my medium."

SPELIUS HAS come out to Idaho for a tune-up. The famous river is the South Fork of the Payette, and the event that has drawn him is the Payette Whitewater Roundup, one of the oldest and most renowned of the regional rodeos. Two years ago, he won it. But there's some new talent this year, as well as some new boat designs, and both factors are likely to push Spelius to his limits. The sport of whitewater rodeo is still young, changing quickly from one year to the next. It's not easy for a 41-year-old legend to keep up.

Spelius himself admits that. "Yeah, life goes on. We're all on this conveyor belt." He takes special pleasure now from the thought of Nolan Ryan pitching a no-hitter at age forty-four, in defiance of time and entropy, but he doesn't delude himself that such defiance can be frequent or lasting. "It's sad that young bodies are wasted on young minds," he says, aware that his own was once wasted exactly that way, "because they've got all the talent, all the assets." Well, not *all* the assets, maybe. The mature Spelius has decades of experience, a sharp analytical intelligence, and a set of ideas defining a savvy new boat.

Two of the best young rodeo stars on the Payette this weekend are Dan Gavere, of Montana, and Lee Bonfiglio, of Oregon. Each of them will be paddling a Pirouette S, the hot new kayak from the other leading manufacturer in the U.S., Perception. The Pirouette S is a lovely multipurpose boat, well suited to river running and recreational slalom as well as rodeo, and it has been a splashy success since Perception introduced it just a few months ago. Most of the top contenders here will be paddling the same model. Spelius, loyal to Dagger, for whom he serves as a traveling ambassador as well as a design consultant, will use one or another of Dagger's boats. But which one?

A few hours before showtime, he still hasn't decided among three different models. One of them is the Crossfire, a sweet number that he helped design. It was a pathbreaking innovation in its time, but that was two years ago, ancient history. Another is the Freefall, a shorter model that might be preferable for this particular hole. A long kayak can be ungainly in a small hole; then again, a short roundish kayak may not deliver the most energized enders. Spelius takes practice in each of the boats, and from where I sit on the bank I can see his brain working as hard as his body. He's wearing his unmistakable pink helmet, with a long tail of satin ribbons dangling flamboyantly off the back, but the expression on his face is sober. His brow is squinched. Driving again and again into the hole, spinning, rising, pirouetting, he seems to be listening with his muscles. If a two-hundred-pound Nordic-looking man in a tasseled pink helmet can appear acutely reflective, he does. Finally he chooses his boat. At the end of the preliminary round he's in seventh place, tied there with a precocious fifteen-year-old boy. The announcer joshingly notes the disparity in their ages. Spelius takes that in good cheer, and he's glad to have made the cut for the finals. Everything can be different in the finals.

But the finals come down to a head-to-head contest between Lee Bonfiglio and another young kayaker, also in a Pirouette S. Dispensing with his paddle, the other fellow cracks off some dazzling barehanded pirouettes, but the judges give greater weight to Bonfiglio's smoothly linked retendos, and so Bonfiglio is the winner. Long before those particular results are announced, a more general truth has become clear: This day belongs to the oncoming generation and to the boat-design shop at Perception.

Spelius is not one to sulk. Besides, he's too busy. Immediately after the finals, as the rodeo officials pack up their tables and the crowd drifts away, as the other competitors carry their boats up the bank, he's back on the water, now in a borrowed Pirouette S. He drives into the hole and executes a nifty retendo. His eyes are alert but undirected. He looks critical, appreciative, discerning, like an oenologist tasting the competition's *grand cru.*

CHRIS SPELIUS has learned what we all learn: that life becomes more complicated, more hectically segmented, as a person passes the age of forty. Not so very long ago he was happily employed as an instructor at the Nantahala Outdoor Center, guiding class after class of kayakers down the Ocoee River in east Tennessee. He spent his free time on the Ocoee as well, ten or eleven months a year, narrowly focused on the sheer physical challenge of the sport, competing in rodeos for the hell of it. Everyone knew him—though most knew him from a distance—as Spe, the big dude with the corn-silk hair and the mile-wide confidence, the charmingly incorrigible lug who gave his motivational exhortations in terms of "studs" and "babes," the Captain Marvel of whitewater. "You know, the year I was totally dominating the Ocoee, I was paddling three hundred days a year," he tells me. "Probably like Gavere is now. Didn't have any other responsibility. Didn't have any reports to file."

"Didn't have a PowerBook," I suggest.

"Didn't have a PowerBook," he agrees.

Nowadays he carries his little black laptop, as well as his paddle bag, through a lot of airports. His life is divided into pieces: running a whitewater business in Chile during the winter, coming back up in spring for the North American season, consulting for Dagger on boat design and showing the Dagger flag at major rodeos in the United States and Japan, serving on the international steering committee that organized the forthcoming Worlds, giving slide presentations and clinics, and returning occasionally to do a guest lecture at Nantahala, where he's thought of as an elder statesman. He keeps up with the details of his far-flung concerns via e-mail. In Idaho this weekend, he has been getting himself online through a pay phone at the same restaurant that offers greasy rib steaks. By 9:30 on the morning after the rodeo, he has already squirted off a two-page memo to the president of Dagger, Joe Pulliam, arguing the case for his hypothetical boat.

The dominance of the Pirouette S in yesterday's competition only confirmed what Spelius already knew: that if the Dagger folks want to remain competitive at the highest level in rodeo, they need to produce a radical new kayak. And they need to do it quickly—in time for the Worlds in October. It doesn't take long to build a kayak if the design specifications and the production mold already exist, but creating one from scratch is another matter. Furthermore, Spelius will need a few weeks to practice with the new boat. This summer of all summers, he'll have to push his business distractions away and focus, like in the old days, on the sheer physical challenge.

"For the last three years, I've gotten away with paddling less than other guys," he says. "Because of my background. But now these kids are coming on so fast. I've gotta get in the water."

By early afternoon he has his answer from Dagger. Joe Pulliam and the other home-office honchos have conferred, anguished briefly, and then flicked on the green light. Yes, okay, despite the fact that they're already overextended with other production and marketing chores, they'll undertake a crash program to build the boat. Not a commercial model, necessarily, but a Kevlar prototype that Spelius can use in the Worlds. Now would he please get his butt back to Tennessee and help them figure out what this new thing should be?

* * *

Water is his medium, Spelius says, and it always has been. He remembers himself as "a water baby." His mother was a Montana state swimming champ fifty-some years ago, competing in backstroke and freestyle, and she seems to have passed both the aquatic gene and the competitive one to her children. She also helped them to discover the satisfactions of free-spirited, outside-the-lanes adventure. During the summer after Chris finished second grade, in Cincinnati, she organized and led the first river trip of his life—aboard a small motorboat, cruising and camping from central Montana back to Cincinnati, along various stretches of the Missouri, the Mississippi, and the Ohio. Chris and his sisters and brother lived the life of Huck Finn, swimming, exploring the barges and the docks that they tied up to, getting themselves in and out of trouble, abiding by only one cardinal rule: Each kid wore a life jacket at all times. Beginning that summer and considering all his raft-guiding and kayaking time, Spelius guesses, he may have spent more total hours in a life jacket than any other 41-year-old person in the world.

All four children were strong swimmers. During summer vacations that the family spent camped at a remote site on Isle Royale in Lake Superior, Chris and his siblings competed to see who could swim out the farthest before exhaustion and chill forced them to turn back. His mother loathed that game, but she had the consolation of knowing that her children were physically confident, and that they were each other's best friends. Later, as an undergraduate at the University of Utah, Chris was still swimming competitively—backstroke and freestyle, like his mother. He also played water polo. Although he didn't star on the Utah swim team—he worked hard but he wasn't supremely talented for that sport, and his body didn't match the ideal type—he did have the advantage of huge lungs. Still later, during a stint with the U.S. Olympic flatwater kayaking team, physiologists at an Olympic training center would tell him that he had the second-largest lung capacity they had ever tested. His other large capacity was for derring-do. After swim practice, he and one teammate would goad each other to make death-defying, out-over-the-tile dives off a second-floor handrail.

His kayaking began at Utah. One winter he took a course in the pool, loved the feel of boat against water, got motivated, and by springtime had mastered the Eskimo roll (that crucial little trick whereby an upside-down kayak is flipped right side up) in almost every conceivable variation: standard roll, offside roll, barehanded roll, one-handed roll. For others, this takes years. Spelius was working on an elbows-only roll before he had ever put his boat on a river.

His running buddy and mutual provocateur during this phase was a small demon-eyed pal named Ken Lagergren. By way of their indoor winter workouts, they had learned all the basics—or so they figured—and now they wanted the world of real rivers. They wanted big water. They wanted wahoo. Along with a third fellow, they charged out to Westwater Canyon on the upper Colorado while it was in flood stage, flowing at 30,000 cubic feet per second. "Here we are, first-year paddlers. Lumber, logs, dead cows coming down the river," Spelius recalls. "It was survival." Days afterward, he could close his eyes and still feel Westwater's momentous surgings, still feel the exhilaration of being violently dumped, engulfed, and then rolling himself back up into daylight. "I love this sport," he thought. "This is my sport. This is it."

The early 1970s. Richard Nixon was president and Dan Gavere was a toddler.

KAYAKING WAS different then. For one thing, the boats that Spelius and Lagergren used were early fiberglass models, primitive by current standards. They were longer, more voluminous, less durable, less ingeniously sculpted, less suited for balletic maneuvers and thumping stunts than the plastic boats of today. Roughly four generations of major design innovation stand between those kayaks and, say, the Pirouette S. For another thing, the level of creative athleticism among kayakers hadn't yet risen far. By analogy with basketball, it was the age of Bob Cousy and black high-top sneakers, before Oscar Robertson had invented the modern jump shot. A small number of paddlers on the East Coast had discovered whitewater slalom, with help from some visiting Europeans, but out on the wild rivers of the West a kayak was a touring vehicle that allowed contemplative folks to make solitary trips down remote canyons. So far as those folks were concerned, Spelius and Lagergren were heroically crazy. They ran Westwater in flood. They ran other harrowing chutes of chocolaty water that had been considered off-limits for sensible boaters. And they paddled their kayaks straight into holes—on purpose. "We were big fish in a little pond," Spelius says. "We thought we were great because we were better than the other people around us."

It wasn't enough. They wanted ultimate wahoo, not just the best of their region, so they went east on a commando adventure and ran the Niagara Gorge—not the falls, no, but the thunderous mile-long canyon below—an escapade that was prohibited by local law, since too many people had been killed. The Niagara Gorge flowed at 120,000 cubic feet per second, with standing waves and monstrous swirls that made all the other famed stretches of North American whitewater, including the Grand Canyon, seem dainty. They both survived, and Spelius, more lucky or more cunning than Lagergren, even eluded the cops. Sometime around then, they fell into the jocular habit of addressing each other as Stud. It was the saga of Stud and Stud, pals and competitors, two human porpoises at large on the waters of America.

Spelius stayed in the East and began teaching at Nantahala. Evidently he made a first impression of cockiness there, exacerbated by his cowboy boots and hat. "People tell the story that I cruise into the dining room, throw my hat down, and say, 'Hi, I'm Chris Spelius and I just ran the Niagara.' " But it's a libelous exaggeration, he says amiably. At a quieter moment, he adds, "I've never thought of myself as a big guy. I'm just my mother's boy, same as always. You know, the same shy kid you were when you were little." There's a misleading discrepancy between the two Spelius personae—the brash Spe whom acquaintances think they know, the more complicated Chris with his hidden depths of sensitivity, wholesomeness, vulnerability, and reflection—that extends back to those early days. His physical presence, so imposing (at least relatively, since most elite kayakers are smallish people), probably contributes to the discrepancy. So does his love of competition, which can be mistaken for a love only of winning. Even the bandying use of his pet word, *stud*, is mainly a goofy motivational tool for goading himself and his students and clients to higher performance, not a reliable gauge of his attitude toward gender relations. And the satin tassels behind his helmet? "People ask me what the ribbons are for. You know, they fly out when you go over a drop, and it's fun. I say, 'That's for a babe on every continent.' But it's bullshit." He follows this admission with a self-effacing smile.

Flatwater sprint kayaking, an Olympic sport, was the next focus of his competitive zeal. He made the U.S. team in 1984, went to Los Angeles, had a rich mixture of thrills and frustrations, and then set flatwater aside. Two other signal events from about the same period were connected more lastingly to his life: He won his first rodeo, and he got his first glimpse of a river called the Futaleufu, on the western slope of the Andes in Chile.

The rodeo victory, at the Eastern Freestyle Championships in New York, derived from his long fascination with the zone of chaotic equilibrium. "I like holes," he says. Back in those days, most boaters didn't. Spelius found that the main impediment to fancy hole-riding was mental. The spins and the rolls weren't actually too hard—not as hard as, say, perfecting a forward stroke for flatwater racing—and the risks weren't usually too dire. "But everyone was so scared of it, and they were so impressed when you went in, that they made it out to be a big deal." He discovered he had an aptitude for this sublime foolishness.

In 1986, he won the Rogue Rodeo in Oregon. In 1990, he won the Dust Bowl Rodeo on the Arkansas River in Oklahoma and took second at the Ocoee in Tennessee. In 1991, he placed first at the Payette, second at the Japanese Whitewater Rodeo, third at the World Rodeo Championships in Wales. In 1992, another first in Oregon, another second in Japan, another second at the Ocoee. But meanwhile Spelius was beginning to spread himself thin. Because of the friendships he'd built with some of the best European paddlers, such as Jan Kellner of Germany and Andy Middleton of Britain, he got involved in planning the forthcoming world championships. Memos and letters from his PowerBook came tweeting out of e-mail linkups around the world. And his own business, Expediciones Chile, made stronger claims on his time and his heart.

His first trip on the Futaleufu had been a solo exploratory run. He expected a sequence of terrific rapids, but he didn't know just *how* terrific (or terrible) they might be, since no one before him had charted or labeled them. He was in roughly the same situation as John Wesley Powell back when Powell's boats first entered the Grand Canyon of the Colorado. Although the Futaleufu isn't quite as big as the Colorado, to a lone kayaker it might as well have been. "I came down through the upper canyon," Spelius says, "through the Inferno—hadn't named the things yet—through the Throne Room, through the Zeta, and came out of that canyon into this . . ." Words momentarily fail him, but the essence comes across: into a scenic paradise after the flush of Hadean excitement. He gazed at a widening valley where another river joins the Futaleufu, with mountain peaks above and a beach of white sand, and he thought: "This is too much. This is the place." Eventually he bought a small piece of land along the river and established the rustic, comfortable camp that serves as base for his guided trips. Now he tells anyone who will listen that the Rio Futaleufu—parts of which are screamingly difficult, other parts friendly enough for intermediates—is the most glorious sampler of whitewater in the world.

There are no announcers, no judges, no bibs, no sponsorship banners, no national teams, and no boat-manufacturer rivalries on the Futaleufu, though there are some world-class holes. The only rodeo is a rodeo of the spirit.

THE NEXT time I see Chris Spelius is on a Monday in October, five days before the start of the World Whitewater Rodeo, which will be held on the Ocoee, a river that he knows well. The Ocoee isn't as spectacular as the Futaleufu, but it's far more practical as venue for a high-profile international competition. As I drive east in a rental car from the Chattanooga airport, the lower slopes of the Smoky Mountains are spackled with autumn colors: maples and sweet gums and sumacs gone red, dogwoods as purple as sirloin, buckeyes and beeches and tulip poplars in yellow. The season hasn't quite reached its crescendo, and the oak leaves are still green. I spot Spelius standing on the riverbank just above Hell's Hole, where the main event will be held.

Several important developments have occurred since July. He has had ear surgery, a bone-cutting procedure to correct a lifetime's worth of water damage that recently became troublesome, and his recuperation has cost him three weeks of practice. In September he turned forty-two. On a more cheerful note, he has gotten his new boat. The hypothetical retendo machine is now real. Proudly, he shows me. It's scarlet and shiny and it looks like the beak of a duckbill platypus.

The wizards at Dagger have produced four of these odd prototypes, in three sizes, the largest of which (and the last to be ready) is for Spelius. The smaller versions will be used by certain other top contenders, one of whom is Marc Lyle, an affable young man with a wrestler's build who had never entered a rodeo before this spring, when he and others discovered that he's a formidable competitor. During the past couple months, Lyle has helped Spelius help Dagger to refine the new design, and the two of them share a warm camaraderie, cemented no doubt by long days of tumbling through Hell's Hole in twin boats that resemble an aquatic monotreme.

Dagger's major competitor, Perception, has also produced a new kayak. This one looks slightly less platypoid, though it incorporates some of the same general features: flat nose, flat tail, narrow-blade silhouette, short overall length. Perception's has been christened the Pirouette SS, and like Dagger's it's an experimental prototype, not available in stores. Dagger's duckbill boat is being called the Transition, a name that may lack the flair of Crossfire or Pirouette but seems especially appropriate when Spelius, the man in transition himself, paddles it. Both of these kayaks are designed for a very particular purpose at a very particular place and time: to do retendo moves in Hell's Hole during this year's Worlds. After that, nobody knows. Nobody at Dagger or Perception can predict whether recreational hackers, like me, would ever want such a boat or be capable of using it. Maybe both models will go into the catalogues, or maybe they'll go onto the junk heap of history. One thing is sure. Each company would dearly love to have bragging rights from a world championship.

I watch Spelius paddle his Transition into Hell's Hole. He surfs, he spins, he drives down for an ender, he rises vertical, he pirouettes, and then with a quick reverse stroke he lands himself back in the hole. As he tries a few further spins and dips, some of which flow smoothly and some of which don't, I see the analytical squinch reappear in his brow. After one small miscue, almost unnoticeable, he gives his head a slight shake. Finally he flips, washes out of the hole upside down, and rights himself with an easy roll. As he climbs out, I ask for an assessment of the boat.

"It's tricky," he says.

* * *

FIVE MILES upstream from Hell's Hole is a Tennessee Valley Authority powerhouse, where the entire Ocoee River passes humblingly through a set of turbines. A half mile below that is a spillway dam and the headgate of a flume that leads to another powerhouse. These features together control the river's flow, like a tap handle controls the flow through a faucet. Just below the dam is the standard put-in for boaters, and from there the river gushes through four miles of delightful whitewater.

It gushes, that is, when the flow is turned on. With the flow off, the Ocoee is an empty channel, cobbled with reddish-brown boulders and punctuated with puddles.

The schedule of releases takes account of recreational whitewater use, though power-demand rhythms are important too. On a typical day, the cutoff might occur around 4:00 P.M. Exceptions to the posted schedule are occasionally made, and at times the invisible TVA rivermeisters seem as perversely generous, or as petulant, as the gods of Olympus. When the flow is suddenly cut at the dam, the lower reaches at first appear unaffected; they continue frothing and splashing until the last slug of water has passed by. That takes almost an hour. But at any given spot—Hell's Hole, for instance—the change happens quickly. The water level drops away, and with it the river's energy and beauty and excitement, like an abrupt tidal ebb.

I SPEND most of the week on the riverbank beside Hell's Hole, watching Spelius and a hundred other competitors practice. It's an impressive selection of the world's best kayakers and canoeists. Some of them are familiar to me as acquaintances or resonant names: Dawn Benner, Nancy Wiley, Becky Weis, Hannah Swayze, and Risa Callaway in the women's division; Dan Gavere, Lee Bonfiglio, Corran Addison, Bob McDonough, and a few others among the men. Jan Kellner is here, the defending world champion. So is Andy Middleton, who placed second at the 1991 Worlds, just between Kellner and Spelius. Shane Benedict of the United States and Shawn Baker of Britain are also contenders. There's a large contingent of plucky Japanese, mostly paddling Crossfires, which means that they're two years behind the technological curve. And from the 1992 U.S. Olympic team in whitewater slalom, Scott Shipley and Eric Jackson, two world-class racers crossing over to test their skills at rodeo.

Shipley and Jackson, I notice, are paddling the two other prototype Transitions. Benedict and Bonfiglio have each been favored with a Pirouette SS, which are also in preciously short supply. Bob McDonough, Perception's master designer of whitewater boats, appears more comfortable in the SS than anyone else—not surprisingly, since he shaped it with his own hands. While McDonough executes a long series of graceful, controlled spins in Hell's Hole, I talk with Kent Ford, a two-time world champion in team slalom. Ford is now retired from serious competition, at the ripe age of thirty-six, and is here shooting a video. The sport of rodeo, he tells me, has changed in recent years.

"It used to be, you'd go into a hole, and twirl your paddle, and get the crowd on its feet—and you'd *win.* Nowadays it's more a matter of sheer paddling skills," he says. "The element of showmanship has gone out of it some." There's less panache now, and more emphasis on a repertoire of technical maneuvers, which might look easy when performed by an extraordinarily fine paddler but which in fact are subtle and difficult. "Like that," says Ford, as someone executes a pirouetting ender back into the hole. "That's a *very* difficult move." He cites Bob McDonough as an exemplum—a great technical paddler whose performances are surpassingly skillful yet so smooth and steady that the degree of difficulty doesn't show to a casual spectator. McDonough has won his share of rodeos in the last three or four years. On the other hand, Ford says prophetically, brink-of-disaster excitement and showmanship can still play a part in judging.

McDonough himself later alerts me to another recent difference in rodeo-paddling on the Ocoee: Hell's Hole has changed. McDonough has paddled the Ocoee for more than a decade, and throughout most of that period, Hell's Hole was nothing. Not a major local attraction, anyway, let alone an international one. It was just a little tuck in the current, incapable of holding a boat or supplying the energy for a fancy set of moves, and it was disconcertingly located just upstream from a bridge piling. People ignored it. The real action was at a rapid called Double Trouble, miles above. But in the mid-1980s the bridge was torn out and a new one built, spanning right over Hell's Hole without obstructive pilings. Then came an epochal flood, several years ago, that rearranged boulders on the river's bed. Double Trouble was permanently ruined. "We lost that one, gained this one," says McDonough. When the flood tide receded, lo and behold, there was Hell's Hole.

Toward the end of the week, I witness a demonstration of the hole's power. It's late afternoon and the eddies on both sides are crowded with competitors, waiting their turns to practice. Intermittently they've had to wait also for recreational traffic, rafts and kayakers passing through. Now along come another three kayakers, intermediates by the look of their strokes, cruising down through the practice area. Two of them slide by on the left. The third, a young man in a green Dancer (once Perception's best rodeo boat, now passé), notices Hell's Hole only at the last second. He makes a frantic effort to evade it. Instead he drops straight in, and it grabs him. He leans on a panicky brace. He tries to scull himself out, flips, manages to roll up, and finds himself still stuck. At that moment the rodeo crowd lets out a collective whoop of appreciation. He's getting a *ride!* He's getting what we're here to get—though God knows he doesn't want it! The poor fellow struggles, flips again, rolls up again looking breathless and desperate, by which time we're all hooting and whistling and cheering him on. He only wants out, but people are hollering: "Go! Yes! Stuff it!" He hates the whole deal, but we're loving it *for* him. Finally the hole releases him on a fortuitous surge and he paddles weakly downstream toward an eddy. Everyone applauds.

SPELIUS devotes part of the week to hard practice, part to stretching and rest, part to diplomacy. Some of the European competitors are understandably concerned about how the American organizers plan to structure the judging. Because Spelius himself wants this to be a friendly and unifying occasion, not a divisive one, he tries to mitigate resentments and broker a compromise. It helps. People trust him. He'd also like to mitigate the sharp rivalry between Dagger and Perception—notwithstanding the role he himself has played in it by advocating the new boat—if he could figure how. Rodeo should be larky, after all, not a high-stakes form of corporate competition. But that's another measure of change. The sport has gotten more sophisticated, more serious, more technical, and now it's in danger of turning pro.

On Wednesday afternoon, as he goes through his regimen of

stretching, I ask what he's focusing on. "Getting my body to feel good," he says. He means: fit, pain-free, strong, and limber. "Getting my body to feel as good as possible on Saturday and Sunday. I *have* to. I've got a little older machine to work with." He laughs. "You know about that," he tells me.

What about the boat? I ask. Still tricky? Better, he answers. Each day a little better. He's delighted with what Dagger has given him, and now he just needs to make his own adjustments.

Meeting again over dinner, we talk for hours about rodeo kayaks, rodeo politics, the long hard summer of his boat-design project, the bonding that he feels with fellow competitors like Kellner and Middleton, time and change, the state of the world, the various transitions that presently loom in his life, the sociopolitical travails of Chile, and another of his heartfelt concerns, the conservation of rivers. The Bio Bio, his second-favorite Chilean river, is being ruined by an unnecessary dam. Maybe it could have been stopped, by way of some orchestrated persuasion and pressure. He describes a small private hope: that rodeo championships like this one might help to establish a global network of expert paddlers, linked via e-mail, keeping each other mutually alert and activated about threats to the world's great rivers. Competitive rodeo is great fun, he says, but what does it really mean? If it can't contribute something to life on the planet, at least in the form of river conservation, what *good* is it?

Later he says: "I've milked the rodeo thing about as long as a person can." He sees that the best of the newcomers are pushing him out. Bonfiglio, Addison, Benedict, Gavere—they're all younger and lighter and quicker. He still has a role to play for Dagger, he hopes, if not as a top rodeo stud then at least as a touring representative, giving slide shows and clinics. But he yearns to spend more of his time in Chile. Bringing in paddlers from all over the world, boarding them at his camp, guiding them on the Futaleufu and a few other rivers, helping them appreciate Chilean culture—these are what give him the most satisfaction now. "That's far more rewarding than rodeo politics."

DRIVING BACK toward my motel, I stop in the dark at Hell's Hole. As my eyes and my ears adjust, I discern two odd facts.

Despite the time of night, the Ocoee is flowing—evidently the invisible rivermeisters have neglected to turn it off. And someone is out there.

It's Marc Lyle, stealing a few hours of additional practice in his Transition. Hell's Hole has been impossibly crowded all summer, worse still this week, and now he's seizing an opportunity to do some blind spins and cartwheels in the foaming dark. From what I can see, he's working the hole well. He's aggressive, very strong, he's moving the boat every which way, and he seems inexhaustible. He's young.

Noticing me, he paddles over to chat. Yeah, he's been here since 7:30. What with the lineup of competitors waiting to practice during daylight, you gotta get while the gettin's good. What time is it now? he asks. When I tell him, he says, "Shit, no wonder I'm tired." Then he goes back out for more.

THE RIVERBED is empty when I settle myself in the Hell's Hole vicinity on Sunday morning, day of the final competition. I've gotten here early, before the crowd, before the announcer, before the river itself. The hole, as a dynamic shape of water, is also not yet present. In its place is a static configuration of jagged, slaty brown rocks. Having gotten a look at those structural underpinnings, I shift my attention away. As I sit scribbling notes, there comes a hiss of moving water against stone. The hiss rises abruptly to a louder, fuller sort of white noise, and then into a foamy roar. I glance up. Hell's Hole in its fullness. The tide has arrived.

Music erupts from loudspeakers. The riverbanks fill with spectators and photographers. The rodeo begins.

Trust me that the climactic details, given what you've read to this point, would seem anticlimactic. Does Chris Spelius win the world championship? No, he does not. Does he come heartbreakingly close? No. Does he acquit himself with grace and dignity, does he cheer his competitors toward heightened performance, does he seem relieved when it's over? Yes, he does. Is he already counting the days until his return to Chile? I think so.

Does anyone else we know make an especially memorable showing? Yes: Marc Lyle, deft as an otter, strong as a wrestler, brings the crowd and the judges alive with a sustained one-minute sequence of linked spins and barrel rolls and cartwheels that some witnesses take to be the single best ride of the competition. This performance draws a wild chorus of cheers, and it puts Lyle into first place at the end of the preliminary round. Everything can be different in the finals, however, and in this case again everything is.

The world rodeo championship in men's kayak, for this particular year, is won by Eric Jackson. The erratic slalom star, known to everyone as E.J., fulfills Kent Ford's intimation about showmanship. The others put up worthy resistance, but E.J. carries a whole suite of advantages: He's vastly talented, he's exciting, he's inured to big pressure, he's a shameless and engaging hotdog in a sport that was born out of hotdoggery, he's quick, light, and young, and not least of all he's paddling a Dagger Transition, the boat that Chris Spelius imagined into being.

Before that occurs, my own time has run out. By late Sunday afternoon the finalists have been announced, and Spelius isn't among them. The head-to-head finals have yet to begin; meanwhile I've got a maddeningly tight connection to my next assignment. I can't stay. So I shake hands with Chris on the bridge, we exchange some words, he gives me the wide smile, and I hurry away. When I glance back, he's already absorbed in conversation with Joe Pulliam, the president of Dagger.

It's a nice parting image of an aging, intelligent athlete. But that's not the one image from this event that resonates in my memory.

THE IMAGE that resonates in my memory comes from several days earlier, on Thursday afternoon. It's just past five o'clock. As usual I've spent my day on the bank beside Hell's Hole, watching hour after hour of brain-rinsing gymnastics made to seem almost easy. I've seen so many pirouetting retendos by now that I almost imagine I could do one. Time to go find some dinner, I decide. The eddies beside Hell's Hole are still crowded when I pull away. Dan Gavere is there, Corran Addison, and a number of others whose zeal and energy aren't measurable on a workaday clock. Maybe they've heard that Marc Lyle was out here last night until ten, blessed by the rivermeisters with a few extra hours.

But the rivermeisters are fickle.

I drive upstream. I savor the autumn foliage. The maples and sumacs are redder than ever, and by now even the oaks are turning. Two miles above Hell's Hole, I round a tight bend through the trees and then come out onto a fresh view of the Ocoee. It's empty. No boats, no water. All I see is a bare channel paved with wet coppery boulders. The river has been turned off. The day's release is finished. The last slug of water has already passed by. Time is fleeting. But the young men, at play below, don't know it.

# *Vortex*

Last year on June 6th the Gallatin River was in flood, surging to its highest level in almost two decades. It was brown as a curry sauce, forceful as a freight train, and cold. It carried tons of sediment and more than a few uprooted trees. It growled quietly as it flowed down the straightaways, and then at each major drop, at each constriction, it roared like howler monkeys in the Amazon night. No one had died in it yet, not so far, not this season, but someone eventually would. Fishing was out of the question. Dilettantish canoeing was highly inadvisable. The beaches were gone, the gravel shoals were gone, the tranquil backwaters had all been overpowered. The beavers had hunkered away as though for an air raid, and the ouzels were temporarily out of business. At an infamous midstream obstruction known as House Rock, the big rock itself was totally submerged for the first time in recent memory, with only a foot-high dome of upwelling water marking its spot like a gravestone. A crowd of onlookers had gathered there, along the left bank, just down from a highway turnout, to gawk at the violent, mesmerizing loveliness of this runaway river. Their point of vantage was good. But ours was still better. We were in kayaks.

We could skim across the current like water striders, though with slightly less security and aplomb. The real water striders were laying up safe, evidently, with the beavers and the ouzels.

Instead of threading our way to the face of House Rock and then leftward beneath its menacing shadow, as usual, we aimed ourselves roughly over the top. Then the current kicked us leftward anyway, and we rode on through. As we flashed past the folks along the bank, several snapped photos, most watched with unenvying curiosity, and one scampered crazily over the boulders yelling, "You're gonna die! You're gonna die!" The yelling seemed rather rude and downbeat until we recognized that it came from a friend of ours, a wry-minded kayaker who'd only stopped at the turnout for a peek before rushing upstream to put his own boat in the water. All he was really warning against was that we'd take ourselves too seriously in front of the cameras.

Just downstream from the choke-point at House Rock is a staircase of minor hazards we call the Boulder Garden. On this day there was no sign of a boulder, all of them buried beneath steep waves and pillows and treacherous backflows of froth. We made our moves through the Garden, cresting the waves, dodging the nastier pockets, ferrying cross-current, getting dumped, rolling up, bracing, hooting with nitwit glee. At the bottom we eddied out, one by one, collecting ourselves for a head-count in a patch of flat water. Here we breathed deeply again. Always above House Rock there comes a surge of adrenaline, a sense of heightened intensity, and always below it—when the Rock and the Garden have been successfully passed, if they have been—there comes a flush of satisfied relief. Today the river was bigger and more brutal than we normally knew it, and we liked that just fine. We savored the satisfied flush. We knew we might never again get to laugh our way over the top of House Rock. We were a squad of overage amateur waterdogs with graying temples and mortgages and professorial jobs, but what the hell, no one watching now, we sat in the eddy trading high fives.

The famously named patches of trouble were all behind us. The emotional crescendo of the day's run—and possibly of the whole whitewater season—was passed. We paddled on downstream toward evening, toward our parked Hondas with Yakima racks and our towels and street clothes, toward late dinners and wives and children and answering machines and cats. We paddled back toward reality as it's known to dry people. The water was still steep and heavy, but not quite so riveting as above. And now along hereabouts there was perhaps, um, a slight lapse of attention for yours truly.

Probably I had started thinking about certain vexatious issues of ecological politics, or maybe about barbecued chicken. Sandy and Mike and Darrick took a sensible line just right of center, and I must have stared vacantly at their backs while I strayed to the left. There were no howler monkeys to warn me. A flat rock, big as a driveway slab, had been engulfed beneath the high water, which set up a ravenous hole just behind it. I fell into this thing. It swallowed me like I was Jonah.

Instantly I was upside down. The water whomped me and snatched at me from all directions. I had one breath of air, gasped in as I'd gone over and good for fifteen or twenty seconds.

That, to my best recollection, is when I began wondering seriously about the subject of fluid dynamics.

A HOLE, in the sense applicable to river-running (as I've mentioned in the previous piece), is essentially a whirlpool laid on its side, with its axis of rotation perpendicular to the main current. It's a cylinder of water and froth that recirculates constantly, in position, like one of those giant spinning brushes at an automatic car wash. For a rough approximation of how it feels to drop into one, you could take a pass through the car wash on your bicycle. Some kayakers know the same sort of feature under other terms: *sousehole, reversal, hydraulic.* Reversal is especially apt because, stuck in the grip of a hole, you'll feel like you've suffered one.

How do these river holes function? According to a renowned authority named William Nealy: "Hydraulics are caused by water passing over an obstacle and creating a recirculating upstream flow below." The velocity of the water dropping over the obstacle, Nealy explains, is far greater than the water velocity below the hole, which creates an excess of piled water with nowhere to go. Gravity pulls some of the piled water back upstream—because upstream in this special case, from the pile to the hole, represents a natural downhill slide—and that water rolls under, down through the hole again, further driving the cycle. "Hydraulics," Nealy adds, "come in an infinite variety and are a source of amusement and/or fear for boaters."

The amusement comes during controlled, intentional hole-riding. Sometimes a kayaker drops into a hydraulic on purpose and surfs sideways there, in a wild and precarious equilibrium, while balancing the boat niftily on its downstream edge. If the hole is steady and relatively benign, various forms of hotdoggery become possible. You can twirl your paddle in one hand while bracing with the other, maybe. You can execute 360-degree spins, maybe. You can toss your paddle away, try to surf and to spin with bare hands, and hope that your buddies are devoted enough to come in on a rescue mission, if necessary. Witless tomfoolery of this sort turns out to be great fun. It also offers a giddy challenge, since any wrong lean in the upstream direction will cause that edge of your boat to catch in the hole's downward thrust, snapping you upside down so quickly you'll think that Shaquille O'Neal has slam-dunked your head into a Maytag.

The fear element, mentioned by William Nealy, is more obvious. Whether you've gone into the hole on purpose or by accident, sometimes it's hard to get out.

Nealy himself is an experienced kayaker, a lucid expositor of the hydrodynamics of rivers, and (most notably) a manic cartoonist. Think of him as the R. Crumb of whitewater, with a high content of practical information as well as disreputable humor, and you'll have the picture. In his vividly illustrated books, such as *Kayak: The Animated Manual of Intermediate and Advanced Whitewater Technique,* he provides vastly more insight into the patterns of flowing water than almost any other author I've ever found. What he is not, though, and doesn't pretend to be, is a scientist.

Fluid dynamics is one of the most complex branches of physics. Why? Just between you and me, it's because liquids and gases can move every which way. Water, to take our present case in point, is a straightforward substance so long as it sits in a beaker, flat as old beer, being measured or weighed or boiled or frozen or otherwise defined by its physical parameters. But as soon as you pour the beaker, as soon as the stuff starts to flow, the physics of water becomes unspeakably complicated. Magnify the beaker into a reservoir (or, preferably, a snow-covered mountain range), then release all that water as a tumbling river: Presto splasho, the complications are virtually infinite. And this near-infinite complexity is in light merely of classical fluid dynamics—let alone the coy wobbles and winks that chaos theory has recently added.

Follow the subject of fluid dynamics into its technical literature, and you come quickly to such concepts as turbulence, laminar flow, viscosity, Reynolds number, boundary layer, shear stress, Bénard cells, and vorticity. Get a loose grip on those few ideas and then it's time to bail out, take my word, because within another page or three you're bound to encounter some nosebleed-inducing mathematics. It's all very erudite but I won't pause to explain even the bit that I comprehend, which is minuscule, since it can't help a man out of a hole.

The fluid dynamicists wouldn't say "hole." They'd say "vortex." Vorticity is a powerful, far-reaching, generalized notion within the study of turbulent fluids. Vortices occur in various manifestations. A whirlpool is one type of vortex. An eddy is also a vortex, though flatter and less energized. Hurricanes and little prairie-wind twisters are vortical. The full-curl horns on a bighorn ram represent vortical growth, as does the shell of a chambered nautilus. There's a genus of water-dwelling protozoa, known as *Vorticella,* on which the cilia are arranged spirally so as to generate tiny vortices that bring the creatures their food.

Just what is a vortex? By one definition (from Hans J. Lugt's *Vortex Flow in Nature and Technology,* a merciful book whole chapters of which can be read without nosebleed), it's merely "the rotating motion of a multitude of material particles around a common center." So a recirculating hole below a flat rock on the Gallatin River is, yes, a vortex.

Now we've learned something. A vortex must have (a) an axis and (b) angular momentum embodied in a multitude of rotating particles. To the casual glance that Gallatin hole seems just a smirk-shaped trough of foam, but if you visualize its various elements, you can begin to see a dynamic orderliness in their interaction. The hole's axis is horizontal, like the overhead brush at the car wash. The hole's angular momentum is produced by the force of the current and the shape of the riverbed. The hole's rotating particles consist of water molecules, plus some air bubbles and sediment, plus a few twigs and swamped insects and other flotsam, plus an eleven-foot-long plastic kayak, plus a wooden paddle. Not every vortex on the Gallatin River is quite so cluttered, but never mind. At the common center around which these particles rotate, you can picture a blue Cooper helmet inside which is my face.

EVEN THE human heart contains vortices, as reported some years ago in the *Journal of Fluid Mechanics.* The article was by B.J. Bellhouse and L. Talbot, of the Department of Engineering Science at Oxford University, and it was dated 1969, although its context extended backward for centuries.

Scientists had long been curious about the elaborate pump-and-valve mechanisms of the heart, so efficient, so durable, so economically designed. Many questions had been answered, but one feature that remained mystifying was the valve between the left ventricle and the aorta. This aortic valve, which closes behind each pulse of blood to prevent backflow into the heart, consists of three flaps of nonmuscular tissue. The *nonmuscular* thing is significant: These flaps move only by reaction (reaction to what? we'll leave that unanswered for a moment), not by exertion. They open freely in the downstream direction; they seal tightly, fitting together in mutual overlap, against any pressure in reverse. Just beyond the flaps, the aortic wall bulges out into three bulbous sinuses, rounded side chambers off the aorta's channel, with one sinus matched to each flap. All right, what causes the valve flaps to close? What function is served by the sinuses? Bellhouse and Talbot used a water-tunnel model to show that the answer to both questions is: vortices. As a pulse of blood surges through the open valve, each sinus scoops in a portion of that flow and sets up a vortical swirl circling back upstream; each vortex exerts pressure against the downstream side of a flap; additional forward flow delivers not just more blood through the aorta but also, tangentially, more force to the sinus vortices; and eventually, at the interval of a heartbeat, those vortices push the valve closed. It's a sweet, simple mechanism based on complex patterns of flow, and most anatomical researchers had totally missed it. One who hadn't missed it, according to Bellhouse and Talbot, was Leonardo da Vinci.

Almost five hundred years ago, during the high Renaissance, more than a century before William Harvey's discovery of the general circulation of the blood, Leonardo had correctly deduced that vortices must form in those aortic sinuses. He even suspected that the vortices must be somehow involved in control of the valve.

Leonardo could be called, with only a slight stretch (and an apology to Leonhard Euler), the godfather of modern fluid dynamics. "He was the first to describe turbulent motion," according to Hans J. Lugt, "to recognize the difference between potential vortex and solid-body rotation, and to study vortical motions in channels and in the wake of obstacles." Intermittently over his whole adult life, when he wasn't busy with painting, or planning a great sculpture that would never be realized, or designing war machines for Lodovico Sforza, or dissecting corpses, or giving careful and loving attention to how birds move in flight, or serving as architect for one oligarch or another, or sketching ideas for a lens grinder or a helicopter or an air-conditioning unit, Leonardo studied the movements of water.

He observed, he experimented, he analyzed. He sprinkled grass seeds onto flowing water and watched to see where the tangles of current carried them. He noticed that seeds dropped close together on the water's surface did not necessarily stay together. Leonardo had what has been called a superhuman quickness of eye, a gift for preternaturally fast and exact visual comprehension, and with that eye he saw finite patterns in the near-infinite fluid complexity. He wrote about what he saw. More importantly, he drew. Almost all of this work was done privately, even secretly, and has survived only in his notebooks. Some of the written material, maybe much of it, is not by any means brilliantly percipient. At one point he bragged that "in these eight pages there are seven hundred and thirty conclusions on water," but a sympathetic scholar sets this brag in perspective by confessing that "the total effect of his writings on water is to my mind rather discouraging." Leonardo was obsessive on the subject. But the obsession didn't always convert well into writing. His drawings are another matter.

Throughout his career he produced images of moving water—diagrams of spout-released flows under varied pressure, plans for canals and drainage projects, sketches of viaducts, landscapes of river valleys, little doodles of whirlpools and river bends and breaking waves. Two sets of water drawings stand out above all others. The first set was done around 1509, when Leonardo was in his late fifties and still living in Milan under the benevolent patronage of the governor, Charles d'Amboise. These drawings are precise, realistic, and cool, obviously informed by Leonardo's many hours of fastidious observation. They depict the patterns of graceful turbulence that water can assume as it flows against an obstacle or over a drop: curls, eddies, foamy pillows, long waves with the crests peeling over, spirals within spirals, all of these shapes layered down upon one another to give a sense of translucent depth. At the time Leonardo drew them, he must still have believed that the world of human and natural phenomena was an edifice of comprehensible orderliness, and that, through his own vast talents and efforts, he himself might comprehend a fair bit of it. In that sense, these drawings express one of the defining themes of the Renaissance itself.

The other set of water drawings, known as the Deluge series, suggests something quite different. These drawings were done around 1514, when the good Milan years had ended and Leonardo was living in Rome—where he found himself neglected, frustrated, unable to get commissioned work, on the outs with the Pope, and overshadowed by Michelangelo and Raphael. He was now in his early sixties and, as a famous self-portrait from that period reveals, ferociously dour. His contemporaries had come to think of him, says one scholar, "less as an artist and more as an old magician whose mind was stocked with terrible secrets about the universe." The Deluge series consists of ten black-chalk drawings in which Leonardo portrayed the destruction of the world by cataclysms of rainstorm and flood. Trees flattened, mountains ripped open, towns pounded to rubble—by force of water. In these powerful drawings, which seem to reflect an intellectual despair as well as a subliminal thirst for revenge, Leonardo converted all his most graceful vortices into a form of Last Judgment.

It appears that he'd lost faith in the human enterprise, including such exertions of rationality as fluid dynamics. But he hadn't lost his respect for water itself.

When I reach William Nealy by telephone—yes, we're back to Nealy now, that other water-obsessed artist—when I phone

William Nealy to ask how he knows so cotton-pickin' much about moving water, he tells me that it all came the hard way. In the course of kayaking, getting dumped, getting thumped. Swimming for his life. Swallowing more than his share of some of America's most thrilling rivers. "Your basic multiple near-death experiences," he says dryly. Did he ever study fluid dynamics? No. Did he ever take a course in hydrology? Never. No, he says, his mode of learning has been strictly haphazard and experiential. Then he answers my next question before even hearing it. "I guess the only hydrology diagrams I've seen that ever made sense to me," Nealy says, "are Leonardo da Vinci's old drawings."

"Which set?" I ask.

SO I WAS stuck in this hole on the Gallatin, and from where I sat it felt like the end of the world. Time seemed to have stopped. Everything had gone dark and swirly. The breath in my lungs turned sour as I held it, held it. I hung there like a dipstick. I remember the sensations quite vividly.

The hole was gargling my boat. I didn't know just which way was up, but I knew that wherever it was, I wasn't. The water's force seemed to come from all sides. It was shockingly strong, and beyond sheer strength, it seemed angry: like a pissed-off Old Testament God, say, or a six-foot-eight mugger. I took one hand off my paddle to grab at my glasses, which were being pulled off my face, and then I almost lost the paddle. I could swear that a voice spoke to me from amid this dark moil: *All right, chump, I want your paddle and your glasses, yeah, and your helmet and your life jacket as a matter of fact, and I want that gold off the back of your front teeth. I want your boat. I want your booties and your shorts and I want all your money. After that, we'll consider your life.* I held onto my pieces of gear. I wondered: What should I do? This was no place to try an Eskimo roll. Even if I did find the surface, I'd only be slam-dunked back under. This was no place to exit my boat and start swimming. As a swimmer, with less buoyancy and less drag, I might cycle around in the hole until nightfall. This was no place to be, period. But it's where I was. So I waited.

I dangled my torso down like a keel, risking a bump on the helmet, taking that chance as tradeoff for another possibility, which would be more welcome. Holes don't like gargling boats with big keels. I knew that somewhere, below the surface, this vortex had angular momentum that could nudge me downstream just enough to escape the recirculation.

Yes. Yes, eventually, after an eternal-seeming stretch of seconds, I felt the turbulence subside, as my boat broke free. Out. Floating downstream. A seed on the current. I rolled, rising back into daylight. I breathed.

When I caught up with Sandy and Mike and Darrick, they looked at me vaguely as though to say: Where have you been?

# *Only Connect*

The man in the green canoe is Dave Foreman, former Goldwater Republican, former hellion, former Marine OCS candidate, former farrier, former Washington lobbyist, former monkey-wrencher, cofounder of Earth First!, and widely regarded as this country's most influentially radical conservationist during those years when Reagan-Bush resource policies cast their dank shadow over the American landscape. His canoe is a Penobscot, riding low with a ten-day supply of provisions. He dips a paddle. He snaps the tab on a beer. He savors the sight of a great blue heron as it takes wing from the riverbank and cranks upward across a pale August sky. "GBH," he says quietly, for the benefit of those gazing elsewhere. Foreman is a protean man with a durable set of principles, a mild private demeanor and a wild public image, a sense of humor. He's also a survivor. Within recent years, he has survived a poisonous spider bite, an FBI raid, a conspiracy indictment, a trial, and hepatitis. Today he's an amiable middle-aged guy in a slouchy camo hat, cruising downriver with binoculars and a bird book. From a crooked pole jammed amid his canoe cargo flies a Jolly Roger bandanna, apt symbol for a piss-and-vinegar insurrectionist who knows the wisdom of not always taking himself seriously. Foreman's not in retirement, God forbid, but he is on vacation.

The river is flat as a griddle, and broad—a great shallow slick of brown water flowing between soft grayish bluffs. It's the Missouri, graceful and solitary here in Lewis-and-Clark country just below Fort Benton, Montana. Ancient cottonwoods, snaggy and unpruned, stand in groves along the bottomland. Pelicans rest in the backwaters. A few Herefords graze on the banks; they stop chewing, at leisurely intervals, and lift up their muzzles to gape. Foreman speaks in low tones across the water, conducting a desultory conversation with a man in another canoe, a red one, some yards away.

The man in the red canoe is Dr. Michael Soulé. Whereas Foreman is bulky, beer-bellied, strong, and could pass for a blacktop contractor, Soulé is lean and slight, with a small professorial beard and the hands of an orthodontist. He's not a stuffy man, far from it, but he chooses his words thoughtfully—even the jokey ones. Out here on the river, he wears a life jacket and sunscreen and soggy sneakers and a straw cowboy hat, beneath which he vaguely resembles the comedian George Carlin cast against type for a western. Among average Americans, who don't read the technical journals of ecology or population genetics or that discipline now known as conservation biology, Soulé's name is not so familiar as Foreman's. But he's eminent within his field, which is the scientific study and preservation of biological diversity. He was the first president of the Society for Conservation Biology, after its founding in 1985. He was the organizer of crucial conferences, the editor of landmark books, the author or coauthor of important papers that continue to serve as load-bearing walls for the whole edifice of conservation-biology theory. He helped to define the concept of minimum viable population, for instance, and to develop a method of assessing the various sorts of jeopardy faced by severely endangered species. His reputation is global, though limited mainly to that global community of field biologists, theoreticians, and resource managers who concern themselves with the abstruse scientific aspects of conservation.

Soulé's canoe is an Oscoda, and it appears to have traveled fewer river miles, taken fewer scratches and thumps than Foreman's. There's no Jolly Roger. But it's carrying its share of the food and the equipment and the beer.

From the put-in at Fort Benton, Foreman and Soulé and their party intend to cover 150 miles—among the wooded islands and past the old ferry landings, through the White Cliffs area, through the Missouri Breaks, all of which is officially Wild and Scenic—to a take-out near the western end of Fort Peck Reservoir. Fort Peck is where this majestically sleepy river turns into a comatose man-made lake. The canoeists could be there in less than a week, if they lean on their paddles, but they aren't in a hurry. Their purpose, so far as I in my nosiness can deduce, is threefold: (1) to enjoy some late-summer peace and some remote landscape; (2) to get to know each other better; (3) to discuss a vastly ambitious, provocative, and visionary new enterprise in which Foreman and Soulé have allied themselves. Their name for that enterprise is the Wildlands Project.

At the heart of the Wildlands Project is the notion of connectivity. This notion proposes that, for scientific reasons, wildlife reserves and parks and wilderness areas and other parcels of wild landscape should be left connected (or be *re*connected by ecological restoration) to each other in a geographical network. Connectivity isn't generally the case in late-twentieth-century America, and less so with every passing year. Instead we have its antithesis, fragmentation, as our forests and prairies and wetlands and natural deserts are inexorably chopped into pieces, leaving discrete and far-flung fragments to stand insularized in an ocean of human impact. The fragments carry names such as Glacier National Park, the Bob Marshall Wilderness, the Little Belt Mountains, the Big Belts, the Crazies, the Bridgers, the Tobacco Roots, the Greater Yellowstone Ecosystem—to cite just one regional set of examples. The bold hope and the quite serious plan of the Wildlands Project is eventually to link those fragments, and other sets, back together.

I've extracted an invitation to share part of this river journey and discussion. As the canoe flotilla moves downstream, I skitter between Foreman and Soulé in a kayak, the craft of choice for gadflies. It's a nifty mode of travel, allowing me to dart in and out of conversations, though it leaves others to carry the baggage and fly the flag.

* * *

Four years ago at a small conclave of Earth First! activists geared toward spiritual reinvigoration and tactical brainstorming, I heard Dave Foreman tell the troops: "Go out and talk to a white pine. Go out on a moonlit night and try to *connect* with the soul of a 200-year-old white pine."

Earth First! is now part of his past, his own sense of tactics has evolved, and his "try to connect" message presently carries a different meaning. The new meaning is more geographical than spiritual, more scientific than mystic. We've got to connect the last fragments of wild landscape to one another, he argues. We've got to link them, by way of corridors that allow fauna and flora to pass back and forth, if we hope to stop the deadly trend of species being lost from those fragments through a process that some scientists call ecosystem decay.

This argument has a distinguished pedigree, as Dave Foreman is well aware. He knows that it comes straight out of a branch of science called island biogeography, which involves much more than islands. As a studious nonscientist who pores through the scientific literature, Foreman knows that island biogeography grew to prominence during the 1960s and 1970s, became the preferred theoretical framework for many ecologists and population biologists, and served as one of the chief sources for the ideas of conservation biology. He knows that ecosystem decay is the process whereby a patch of habitat—if it's too small and too thoroughly insularized—loses species as though spontaneously, the way a lump of uranium loses neutrons. For instance: The Bridger Mountains, once they were insularized, lost their population of grizzly bears. The Crazy Mountains, once insularized, lost their grizzlies. The Greater Yellowstone Ecosystem, a much larger area, retains a grizzly population to the present—but even that population may eventually be lost. Each loss of this sort can result from subtle causes involving only the mischances of demography, genetics, and environmental fluctuation, whether or not the species is protected from hunting. Those subtle causes have been elucidated over the past thirty years in the journals of theoretical biology. The whole subject of ecosystem decay, and of its relation to the sizes and patterns of parks and reserves, has been pondered exhaustively by the experts while remaining completely obscure to almost everyone else. Foreman is an exception. During his gradual metamorphosis from eco-radical firebrand to conservation-biology wonk, he has read all about it—from the early works, such as Robert MacArthur and Edward O. Wilson's *The Theory of Island Biogeography*, to the later papers of Michael Soulé.

The MacArthur-Wilson theory, as interpreted latterly by conservation biologists, implies that a habitat patch can retain its full complement of species *if* it continues to receive immigrants (wandering individuals) from other patches. How can such immigration be fostered? One possible way is with habitat corridors. Those corridors may be narrow, encompassing no great amount of area, but as long as they provide travel routes for restless animals and dispersing plant seeds, they can accomplish a disproportional measure of good. So says one school of thought, anyway. An opposing school notes that the same corridors might have negative consequences that outweigh any positive ones, such as facilitating the spread of diseases, fires, or other forms of catastrophe that might wipe out a population of creatures. And whether the corridors would be cost-effective, in terms of the financial and political capital required to establish them, is also a much-argued question. Although the anticorridorists include some respected scientists (such as Daniel Simberloff, a brilliantly argumentative ecologist who has served on the board of the Nature Conservancy), Michael Soulé's own research and reflection have landed him in the procorridor school.

The potential role of corridors for preventing ecosystem decay was suggested as early as 1975. In the years since, there has been a great deal of theoretical refinement and debate, though not much field-testing or application. "Conservation biology, as long as it's talked about in abstract terms, remains just that," Foreman tells me. He means: It remains emptily abstract. "What we've got to do is make it real."

I ask Michael Soulé how he came to be involved in the project. His answer surprises me slightly. "I wrote to Dave and I said, 'What's the next step?' Because conservation had come to a dead end. Nobody had any vision. And the obvious thing was connectivity."

Foreman himself was surprised. He showed Soulé's letter to one of his current coconspirators and crowed, "Look, somebody *agrees* with us."

* * *

In a ninety-page document that serves as both manifesto and blueprint, Foreman and Soulé and a few other collaborators have offered a short "Mission Statement" describing the essence of their enterprise. "The mission of The Wildlands Project is to help protect and restore the ecological richness and native biodiversity of North America through the establishment of a connected system of reserves." Healing the landscape, they say, will entail "reconnecting its parts so that vital flows can be renewed." From there they continue in more exuberant language. "Our vision is simple: we live for the day when Grizzlies in Chihuahua have an unbroken connection to Grizzlies in Alaska; when Gray Wolf populations are continuous from New Mexico to Greenland; when vast unbroken forests and flowing plains again thrive and support pre-Columbian populations of plants and animals; when humans dwell with respect, harmony, and affection for the land; when we come to live no longer as strangers and aliens on this continent." Apart from the idealism of that vision, which is admirable, the crucial content lies in the words *unbroken, continuous,* and *connection.*

What it brings to my mind, besides theories of island biogeography, is a certain famous passage from E.M. Forster's novel *Howards End.* Forster's heroine, Margaret Schlegel, was pondering what she might say that could vivify the soul of her obdurate, mean, myopic, business-obsessed, yet righteous and well-intending future husband. At the moment, it didn't seem to her a difficult question. "Only connect! That was the whole of her sermon. Only connect the prose and the passion, and both will be exalted, and human love will be seen at its height. Live in fragments no longer." But the man, Henry Wilcox, was incapable of comprehending her and incapable of making vital connections, either between the two sides of his own nature or between himself and another human—at least until his life had been shattered to still smaller and meaner fragments.

Foreman has given me a copy of this document—not *Howards End,* no, but the Wildlands manifesto—and I carry it in my drybag as we float the river. I've read the mission statement, and Foreman's own introductory note, and a wise essay by Soulé on the theme of incremental consensus-building and long-term patience; and I've looked over the maps. These maps represent specific but tentative proposals from working groups allied to the Wildlands Project in various regions—a hypothetical sketch of reserves and connecting corridors for the southern Appalachians, another for the Adirondacks, another for Florida, still another for the northern Rockies. I've closely examined that last map, the one covering those fragments of wild landscape with which I happen to be personally familiar. I've seen the large dark ovals, indicating park-and-wilderness aggregations as they already exist, and the gray swaths drawn to represent corridors, or proposed corridors, passing fluidly down the page like runnels of chokecherry syrup. I've traced those runnels against the map of Montana in my head, and the fact isn't lost on me that they flow over the Little Belt Mountains and the Crazies and the Big Belts and the Bridgers and the few tiny towns in between, crossing small county roads and ranches and two different stretches of interstate highway, presenting the dreamlike ideal of a restored biogeographic connection between the Glacier Park complex and Greater Yellowstone. How will that dream be realized? How will the landscape be transformed? How will the highways be transcended? My first impression of this grand design has been that it's scientifically sound, ethically compelling, and politically impossible.

Now I paddle over and put that reaction to Foreman. Isn't his project hopelessly unrealistic? Does it have any practical viability in the real world of congressional turf battles and economic stress? Will it ever be tolerable to the hardworking, vote-casting folks who are already desperately concerned for their prerogatives to continue cutting timber or pasturing cows or driving their dirt bikes into the backcountry?

"I'm sort of a practical guy," Foreman tells me. Yes, it's an audacious proposal, he agrees, and he doesn't expect quick victories, but he does hope to turn that proposal into the focal point of the debate over how we should—and *can*—preserve biodiversity on this continent. "I think it's *very* practical." There's a parallel here, I realize, with the guerrilla-theater radicalism of Earth First! Foreman himself, the old stage-stomping, fist-waving agitator, always understood clearly (as many of his followers didn't) that the real usefulness of Earth First! was to move one flank of the debate out to extremity, not because extremity itself would ever triumph but because the middle-ground position would thereby also be pulled in that direction.

"Our job is not to operate within the bounds of political reality," he tells me now, as the river carries us sweetly along. "Our job is to *change* political reality."

We camp beneath an old crabbed cottonwood on the right bank and, after dinner, after dark, the discussion continues. Foreman smokes a cigar, partly to keep the mosquitoes away, partly because he likes the taste, partly no doubt because he finds cigars appealingly retrograde. Soulé and I content ourselves with chocolate-chip cookies. Somewhere across the river, coyotes hoot. I sit on an ammo box, listening carefully to what Foreman and Soulé say about conservation biology, the science of jeopardized populations, and politics, the art of the possible.

Just how much *is* possible? I wonder. The Wildlands manifesto proposes a three-category schema of ecosystem conservation, consisting of core reserves (which would be strictly managed for the protection of biological diversity), buffer zones (multiple-use lands surrounding the core reserves, which would constitute at least marginal habitat, insulating the reserves from intensive land-uses as practiced on the terrain beyond), and corridors (for that crucial connectivity). One step toward realization of the schema might be to establish a Wildlands Recovery Corps as a new branch of the Forest Service. This corps of restoration ecologists and landscape engineers would focus on the potential corridor lands, instituting soil-stabilization measures to mitigate erosion, closing unneeded roads, doing whatever possible to nurse back native vegetation and faunal diversity. The notion of just such a Wildlands Recovery Corps has already been included within one bill introduced in Congress. But that bill has gone nowhere. Judged in biological terms, the WRC in particular and the Wildlands Project in general make excellent sense. The big problem is that too many local people—ranchers, timber-industry workers, off-road-vehicle enthusiasts, inveterate government-haters, and salt-of-the-earth xenophobes, as well as most of the region's congressional delegates—seem to find the very notion of federally mandated landscape reconnection so loathsome and scary.

From my ammo-box chair, I ask Soulé: Won't the Wildlands proposal, with its breathtaking scope, exacerbate the polarity that already makes America's conservation debate seem like a class war?

"I know your question was rhetorical," Soulé answers politely. "But the same could be said of the Emancipation Proclamation. Doesn't it exacerbate polarity? Yes it does. Anytime you extend liberty to another group, it exacerbates polarity. Even if the new group are animals."

And so on. Skeptical questions, thoughtful answers, no sense of easy resolution. The mosquitoes continue feeding until late.

AT THE end of the second day, exhausted and eager to make camp, we come to a fork in the channel. The river map tells us that this is Boggs Island, about forty miles downstream from Fort Benton. It's public land, nicely forested, and it may be our last decent chance today for a campsite. A question arises: Which side of the island should we scout? The question is deemed unanswerable and so, before I know it, Foreman's green canoe has cruised off down the right channel, Soulé's red canoe has disappeared to the left. The whole party splits. Not foreseeing that it makes any difference, I swing my kayak to the right and follow Foreman.

The island is big. The bank along our channel is steep, with no convenient inlet for beaching boats or unloading gear. Foreman and I climb up and find a campsite, passable though not ideal. The undergrowth is thick. Someone complains about poison ivy. Someone else notes that maybe Soulé and his group have found something better. Do we have a plan to rendezvous and confer? We don't. The island is too wide and brush-choked to cross on foot, and Soulé doesn't answer our shouts. So I volunteer to paddle back upstream in the kayak, drop around into the other channel, and find him.

Paddling against this gentle current is like climbing a rope, hand over hand, stroke after stroke. I get almost to the nose of the island before fast shallow water stops me. There I pull my kayak up the bank and begin dragging it overland toward the other channel. This proves more difficult than I expected: heavy thicket, no trail, deadfalls, thistles and thorns, a forty-pound boat with a propensity for snagging itself on branches, and I'm in a pair of neoprene mukluks. Finally I stumble out to the other channel and set off on the current again. I paddle down the length of the island, finding Soulé and his group at the very bottom. They haven't spotted a campsite, no, though they have enjoyed watching a beaver. Okay, I say, can they pull their canoes upstream along that side for a few hundred yards, to where Foreman is waiting? Not hardly, they say, and call my attention to the fact that the bank is impossibly brushy and steep. Then can they *paddle* these barges upstream? Not hardly. So I set off alone, upstream again, to complete my circumnavigation of the island and report back to Foreman. Before long I meet his group, aboard their canoes and floating down. Now I'm getting dizzy. In the interval, they have made a decision: On grounds of poison ivy or fractured communications or some other ineffable factor, Boggs Island is not the place.

At the bottom of the island, we all finally reunite. Soulé's group agrees: no stopping here. It'll be better to go on and, in what remains of daylight, find a spot somewhere downstream. They all paddle away.

I linger behind, adrift on the slow water. My arms are tired and I feel slightly perplexed, contemplating both the small chore of choosing a campsite and the huge challenge of persuading America to accept Wildlands.

The Wildlands Project has great merit. But it will never succeed and probably shouldn't, I suspect, if the effort to realize it becomes an us-versus-them political battle, since us-versus-them is just another form of fragmentation. It can only succeed through patient persuasion, as Soulé warned in his wise essay; it will also require a system of positive incentives for the folks who would be directly affected, and a large measure of (the one thing that Earth First! considered anathema) compromise. Drawing corridors on a map is easy. The hard connections to make are those between aggrieved humans on one side of an issue and aggrieved humans on the other side.

Within a few minutes, Foreman and Soulé have pulled far out ahead and all I can see are dots on the water, one greenish, the other red. I wish them good luck. It's a long, long way to Fort Peck.

# *Grabbing the Loop*

At the southern end of the Andes, coming out of the highlands of Argentina and down across southwestern Chile, flows a little-known river called the Futaleufu. From a continental perspective it's just a tiny hydrological squiggle that doesn't show up on the less-detailed maps. Even the people of Chile have scarcely heard of it. The river is short, about fifty miles in length, and the valley through which it winds is remote. So you might take this for a backwater place. But no, the waters of the Futaleufu are decidedly forward.

Chile itself is a land of climatic and topographical extremes—tilted steeply east-west from the Andean cordillera to the sea and, along its north-south axis, ranging from the breathless aridity of the Atacama Desert to the glaciers and cold storms of Chilean Patagonia. About five hundred miles south of its capital, Santiago, is the Región de los Lagos, the Lake District, a zone of snow-capped volcanoes, verdant national parks, glacial lakes left behind by the retreat of the Pleistocene ice, gentle pastures, fertile fields of grain, timber and fishery operations, and an old-fashioned agricultural economy, all of which cause urban Chileans to view it with roughly the same ambivalent attitude—romantic yearning vitiated by condescension—that New Yorkers direct toward Montana. The *huaso,* country bumpkin, is a half-affectionate Chilean stereotype often associated with this region. Toward the south end of the Lake District, beyond where the vacationers from Santiago and the tourists from Miami generally go, beyond the provincial capital of Puerto Montt, you come into a harder, more edgy realm that feels genuinely like frontier. From there, catch a boat or a small plane onward to the little port town of Chaitén, then start driving into the mountains. After a few hours of bumping along, with the valley walls rising on each side, you'll enter the thrall of the Futaleufu.

From the Pacific coast to the Argentine border, there are just a few villages along that valley, and the road is unpaved. Campesino homesteads are scattered sparsely, many reachable only along dirt paths. Heavy loads travel the paths by oxcart, and the personal vehicle of choice is a horse. Fences and gates are hand-hewn from split logs. The little farms are hardscrabble, low tech, and diversified. The people face the same life-and-death struggles faced by country people in harsh climates everywhere—keeping the crops and the animals alive despite drought or blizzard, stretching stored food through the winter, enduring childbirth and disease with little or no doctoring, doing the sort of hard, risky labor that turns young bodies quickly old. The landscape they inhabit is graced with jagged volcanic peaks, forested mountain slopes, Andean condors soaring on thermals above the deep slot of the river, and a sense of time gently arrested in a matrix of pioneer virtues. Central to everything, like a big moody god, is the river.

The Futaleufu has its source in Argentina, draining out of the shadows of the majestic, sequoia-like *alerce* trees of Parque Nacional Los Alerces. Crossing the border, it picks up tributaries, speed, and heft. By the time it passes near the village that shares its name, it's too deep to ford and too fast for a ferry. Local folk treat it respectfully, staying out of its grasp, whereas thrill-hungry river rats from three continents come to toy with it. The name itself, Futaleufu, reportedly means "big river" in one of the languages of the Mapuche Indians. In certain stretches the Futaleufu is so limpid and sleek that, though sliding along swiftly, it looks like a slab of buffed jade. But don't be fooled. Elsewhere, around the next bend, it explodes fearsomely into foam.

Some knowledgeable people will tell you that the Futaleufu is the greatest whitewater river in the world. That's what they've told me, and that's why I'm here. God help the middle-aged journalist who thinks he can handle a kayak.

Of course *greatest* is an insupportable adjective as applied to rivers, which are infinitely various. If the Futaleufu is in some sense superlative, is the Zambezi of southern Africa secondary? Not hardly. What about the Colorado where it roars through the Grand Canyon, or the upper Pacuare in Costa Rica, or Devil's Gorge on the Susitna in Alaska? Each of those rivers has its own unique balance of menace and charm, capable of gladdening any traveler who hankers for a soul-rinsing immersion in wild water. Whether the balance in a given case leans toward the sheer difficulty of the rapids, or the grandeur of the scenery, or the exotic flavor of the locale, never mind. My point is merely this: The Futaleufu belongs in that class.

It has the difficulty, it has the grandeur, it has the flavor. It has rock walls and dense mossy forests, meadows punctuated by clumps of bamboo, snowy mountains, cool breezes, solitude along its scarier reaches, friendly people on the byways round about. It has blasting summer sunshine in January. It has lanky black cormorants that fly arrow-like paths upstream while you paddle down, as though they don't want to watch what you're getting into. It has ten-pound trout, and if you handle your boat as ineptly as I occasionally handle mine, you might spend some eternal moments among them.

It has a rapid called Terminator.

THREE CLUSTERED volcanic spires loom over the Futaleufu near its confluence with a smaller and colder stream, the Rio Azul. The spires are known as Tres Monjas, the three nuns. They don't look like sisters of mercy, but they do add a lofty, cold beauty. There at the confluence with the Azul, on a small spit of land, sits Campo Tres Monjas. It's a quiet little outpost for kayakers, run by the world-renowned river athlete Chris Spelius.

Spelius, tall and blond, with wide shoulders and a square jaw, is the same complicated fellow whose fortunes I followed (see "Time and Tide on the Ocoee River") through a whitewater rodeo world championship. Though the family name is Greek, he looks like a Nordic heavyweight who might have beaten up on Sylvester Stallone in one of the *Rocky* films. As a former Olympic team member in flatwater kayaking and a granddaddy personage (mid-forties, and still competitive) on the rodeo circuit, he's famous on rivers all over the world—famous for his brashness as well as his skills. But the brashness has always been just a public facade for a sensitive and intelligent private man, and in recent years he seems altogether mellower. Probably his life in southern Chile has been part of the reason.

Finished with the flatwater racing grind, he went to Chile on a winter getaway in the mid-1980s and paddled the Bio Bio, in those days still the most celebrated whitewater run in South America. There on the Bio Bio, Spelius heard talk of another river, far to the south, that might be even more worthy of celebration: the Futaleufu. Two other American paddlers, Phil DeRiemer and Lars Holbek, had recently made what was probably the first kayak descent, and their take on it, as Spelius remembers, was "Oh, man, it's in a different league."

Spelius decided to see for himself. He ran the Futaleufu solo, carrying little more than a sleeping bag and a toothbrush. Each evening he beached his kayak, walked up to a campesino's house, knocked on the door, and—in his helmet and life jacket, looking like a giant otherworldly visitor—asked sweetly in broken Spanish if he could sleep in the barn. By day he paddled his way through a succession of thunderously difficult rapids, which are now legendary in whitewater lore under names such as Zeta, Inferno Canyon, Throne Room. Just a few miles below the tumult of Throne Room, Spelius encountered a lovely stretch with a very different feel. An aquamarine tributary joined the main river there, with a white sandy beach along the bank and three volcanic spires looming above. The little beach marked an idyllic refuge, a sort of hurricane's eye, near the middle of what was clearly the most bodacious river he'd ever seen. "This is the place," he recalls thinking. "I've got to figure out a way to live here."

At that moment, the prospect of seeing this idyllic refuge buried beneath a hydroelectric reservoir was hardly imaginable.

In the mid-1980s, even the celebrated Bio Bio seemed to have a free-flowing future. The very idea of damming a world-renowned Chilean river—entombing its human culture and its natural wonders beneath impounded water—was just a distant dark cloud in the west, moved by a distant and invisible wind that some people call "progress."

AFTER MANY more trips and a slow process of building friendships, Spelius became not just a landowner in the Futaleufu valley but a real neighbor. The white sandy beach, along with a few acres of wooded land behind it, is now the site of his Campo Tres Monjas. The camp is a simple facility—a mess hall with a fine Chilean cook, an outhouse, a wood-heated shower and sauna, space enough for a dozen scattered tents—because Spelius hasn't wanted to erect any superfluous buildings that would detract from the majesty of the river. In addition to that reason, there may be another—wanting not to give his kayaker clients a perilously false sense of ease. Downstream from Campo Tres Monjas, after all, lie more thunderously difficult rapids.

The most formidable of those lower rapids is a certain thousand-yard gauntlet, a prolonged class-five travail compounded of crashing waves and lurking rocks and a blind drop over a six-foot-high ledge, roiling eddy lines, surprise currents pushing every which way, pour-overs where you don't want to get poured, roaring water here, roaring water there, and amid it all a few hideous holes, some large enough to bite hold of a kayak and swallow it the way a coyote swallows a mouse. An impetuous paddler who heads straight down the gullet of this rapid might cover the whole thousand-yard stretch in two minutes, or then again, might never show up at the bottom at all.

About ten years ago, roughly when Spelius himself was discovering the river, an experienced rafter with major expeditionary credentials, Steve Currey, tried to lead a group through here. It would have been a notable feat—first descent of the lower river by raft. But Currey and company aborted their trip at the thousand-yard gauntlet, after dropping a raft into one of the holes and watching it get tumbled unrelentingly there for half an hour. Since this was their point of termination, the rapid became known as Terminator.

* * *

ALL RIGHT, time to get wet. We climb into our kayaks and launch from the sandy beach, peeling away one by one like fighter jets rolling out of formation. The current is powerful and quick even here, where the water is nearly flat. It's even more powerful than usual, because a rainy period just preceding our arrival has made the river exceptionally high.

This will be our second day on the water, the first having passed without untoward drama. There are ten of us, with two guides, and from yesterday's experience I know that I'm traveling with a fast crowd. The strongest and steadiest paddlers—John from Florida, Skip from Maryland, Dr. Dan from North Carolina, and my pal Mike Garcia from Montana, among others—are very damn strong and steady indeed. Skip, a magazine photographer who's here on assignment with me, is even capable of taking one-handed shots with his waterproof camera as we bob through the ten-foot-high waves. "David, turn around and smile!" he hollers, while my face is bloodless with concentration. Our guides are Ken Kastorff and Mike Hipsher, two of the preeminent guardian angels in the world of whitewater. Kastorff, when he's not helping Spelius or guiding his own trips in Costa Rica, is one of the best kayak teachers in America; Hipsher is a former world champion in wildwater racing, which entails running serious rapids in a sleek, unstable boat at top speed. The other paying clients are vigorous souls, excellent kayakers, several of whom have decided that since judgment is the better part of river running, they prefer the footpath detour around Terminator to the option of paddling through it.

And me? I'm a middle-aged fool with a desk job, who now finds himself parked in a small surging eddy just past the no-turning-back point at the top of Terminator. Yesterday I blundered through here without grievous mishap. Having plunged over the big ledge drop and felt myself shoved suddenly rightward, into the maw of disorder, I flipped upside down, rolled upright again, and paddled breathlessly out the bottom. Today I hope to do better. I have a mild case of jitters. If I were smarter or more prescient, I would be terrified.

At first we pick our way cautiously downstream along a rock wall on the left, eddy to eddy, dodging into the current and out again, with Kastorff leading. This whitewater strategy is called a sneak. Halfway down, though, we gather up in an eddy beyond which there can be no further sneaking. Now comes the hard part, and from here each of us will go alone. Sitting relaxed in his boat, delivering stern counsel from behind his zitzy sunglasses, Kastorff reminds us what to expect and what to do—or what to try to do, anyway. See the horizon line thirty yards down? he says. That's the big ledge drop. Take it at a point about eight feet from the left bank, but be angled toward the right. As soon as you've landed, stroke like hell back to the center. Avoid the nasty hole on the left, brace through the exploding waves, don't go *too* far right for God's sake, and be alert for the large pour-over rock farther down. Everybody got it? Duh, yuh-yes, we say. All right, Kastorff says, now watch for my signal from below. Then he toodles off downriver, disappearing over the ledge like a memory of confidence and safety.

I wait my turn. I splash myself in the face, rinsing the fog off my glasses. I watch for Kastorff's signal. Then I go.

I launch my boat over the ledge on a nice rightward angle and, landing, make my move toward the center. But as I drop into the maw, a wave arches up and smacks my chest, high and hard like a blitzing linebacker, knocking me dippy. Now I am upside down in the unholy chaos at the heart of Terminator, holding my breath and waiting for a good opportunity to roll. What I haven't yet realized is that there won't be any good opportunities, not in this stretch of water, and that I ought to seize a bad opportunity before I use up my air.

Finally I try. One roll attempt, unsuccessful. Another attempt, unsuccessful. The water is alive with turbulence, swirling furiously all around my body, snatching at my glasses and my paddle, refusing to let me pop up. The river seems godawful belligerent. Was it something I said? A third roll attempt raises me upright, for an instant, until another breaking wave knocks me back under. I make one more try: a weak roll, a moment's precarious balance, now if I can just catch a little breath, and sit forward, and plant a decent stroke, I might extricate . . . —at which point I drop backwards over the large pour-over rock, landing upside down in a sucking eddy. I lose my grip on the paddle, and it's gone.

Gaa. I've got no air, no stick, no strength. No recourse but to abandon ship. So I do what all kayakers loathe doing. I yank up the spray skirt that seals me into my cockpit, push myself out of the boat, and swim. This is a form of ignominy, but with the ignominy comes oxygen.

I grab instinctively for a small nylon strap—the grab loop, it's called—that hangs knotted through the nose of my kayak. I cling to the overturned hull while gasping for breath. And then, as the boat eases itself out of the sucking eddy, back into the current, I ride along. Not far below, Kastorff paddles over to make the rescue.

I've lost one river shoe, and the paddle, and a bit of face. The shoe and the paddle are replaceable; the face is not crucial for a doddering writer with no pretensions to whitewater mastery. All considered, then, I drag myself onto the bank feeling that it wasn't such a terrible experience.

But maybe I'm wrong. Late the next evening another paddling pal from Montana, Kevin Kelleher, confides to me: "That was the scariest thing I've ever seen."

Surprised, I ask why. Because when you swam to the surface, Kevin says, the suck of the eddy pulled you upstream at once. If you hadn't grabbed that loop as you swept past the nose of your boat, he says, you would have gone back up into the hole and, buddy, you would have stayed there. He felt desperately helpless as a spectator, Kevin tells me, during that moment when it looked like I might not come out alive.

RIVERS CAN die too. Rivers are animate, in their way. They move. They breathe. Every river, except the most hopelessly throttled or poisoned, constitutes a dynamic matrix for an elaborate, organismic network of biological relationships.

To drown a river beneath its own impounded water, by damming, is to kill what it was and to settle for something else. When the damming happens without good reason—simply because electricity is a product and products can be sold—then it's a tragedy of diminishment for the whole planet, a loss of one more wild thing, leaving Earth just a little flatter and tamer and simpler and uglier than before. Already this has begun happening with the Bio Bio, the Futaleufu's sibling to the north. By early 1994, the struggle to protect the Bio Bio from being dammed was a famous international battle that had been famously lost.

The Futaleufu, on the other hand, was still safe and obscure. Or so it seemed, until a private utility company called ENDESA (the same powerful corporation damming the Bio Bio) quietly filed a claim for water rights to the Futaleufu, with a proposal for multiple-dam development that would put most of the valley's agricultural land, most of its homesteads, most of its riverine ecosystem, most of its grace, most of its heart, and all of its great whitewater under a pair of reservoirs. The ENDESA scheme would even bury the road, severing the valley's tenuous connection with the rest of Chile.

The people of the Futaleufu, including Chris Spelius and his campesino neighbors, only caught wind of ENDESA's move in late March of 1994, with barely enough time to file a counterclaim. Immediately they held a meeting, formed a defense committee on behalf of the river and its ecosystem (under the acronym CODDERFU), passed the hat for expenses, and sent two representatives to Puerto Montt to meet with a lawyer. They had grabbed the loop of this issue just in time, escaped from the suck of political docility, and now they were swimming in big water.

The first CODDERFU meeting was followed by many more, as concern spread and resolve stiffened throughout the valley. The local spokesmen for CODDERFU declared that the Futaleufu and its tributaries "should be preserved for outdoor recreation, for agriculture, and for tourism, which is their true value, as a National Park." It was also suggested that a reasonable alternative to damming the Futaleufu could be found—possibly by placing a hydroelectric plant at the outflow of Lago Espolon, a natural lake with significant potential for electrical generation. Then the whole issue, having been raised and polarized, passed into the opaque innards of Chilean bureaucracy, to be decided who-knows-when, on who-knows-what economic or political grounds.

Meanwhile, Spelius had sat down at his laptop and typed out an e-mail message to his boating friends around the world. "To dam *this river* would be like blowing up the Sistine Chapel," he wrote. From a strictly economic angle, it made no sense, there being more to be gained from long-term tourism than from a short-term flow of construction money. Furthermore: "The local populace does not want this project." He could say that with assurance, having spent many hours hearing them out. "This news has changed the priorities of my life," he added. Spelius was now a respected adviser to CODDERFU, and a key fund-raiser, though remaining careful that his own voice was only one among many. The international river cowboy, the otherworldly visitor, had become a conservationist and a community pillar.

THERE ARE many cogent reasons why the world needs the Futaleufu River more than the world needs another dam project, generating bargain-rate electricity to be sold and, largely, wasted. Whitewater thrills for globetrotting kayakers and rafters are (notwithstanding the tourist revenues they bring) certainly the least of those reasons. Others range from the aesthetic to the ecological to the economic. The particulars of each are complicated, and I won't expatiate on them here, but those reasons lie embedded within the general concern I've already voiced: Must we continue making our world everywhere flatter and tamer and simpler and uglier? Aren't we ingenious enough, as a species, that we can think our way beyond the imperatives of unending population growth, invidious concentration of wealth, and landscape transformation, which "progress" so mindlessly implies?

My own preferred argument for preserving the Futaleufu, based on the moments spent contemplating my mortality in its clutches, is this: Humanity badly needs things that are big and fearsome and homicidally wild. Counterintuitive as it may seem, we need to preserve those few remaining beasts, places, and forces of nature capable of murdering us with sublime indifference. We need the tiger, *Panthera tigris,* and the saltwater crocodile, *Crocodylus porosus,* and the grizzly bear, *Ursus arctos,* and the Komodo dragon, *Varanus komodoensis,* for exactly the same reason we need the Futaleufu River: to remind us that *Homo sapiens* isn't the unassailable zenith of all existence. We need these awesome entities because they give us perspective. They testify that God, in some sense or another, might not be dead after all.

Still, the arguments that will decide the fate of this river involve not what the *world* needs but what the citizens of Chile feel they need. Do they need more electricity and a short-term construction boom here or there? Do they need spectacular wild places within their own country, for their own spiritual refreshment and edification? Do they need to keep faith with the pioneer virtues? As an interested outsider, I'm more than a little curious about those questions. So on the days when I'm not trying to drown myself, I get Spelius to take me calling on some of his neighbors.

One is Don Mario Toro, a grave, laconic man in his late fifties, whose house and corrals lie just across the Rio Azul from Campo Tres Monjas. Don Mario, in a wool sweater and brown beret, is shearing sheep when we arrive. He sets down his clippers. After proper courtesies and an explanation of why I'm there, I ask Don Mario a few gentle but nosy questions about his life in the Futaleufu valley. Spelius translates, while I watch the wrinkles tighten around Don Mario's dark eyes.

It was in 1944, Don Mario says, that his family spent their first winter here. His father had brought them down from a farm near the Bio Bio. Don Mario doesn't say just what drew his father to this harsher place, but that's easily guessed: the boundless promise of the frontier. All their possessions were carried on one horse. The first winter, he says, was a bad one—rain, cold, snow. Their animals died, all except one, the horse. They didn't yet have a barn. So they made a jacket of leather for the horse, otherwise it too would have died. Again in 1953 a bad winter, the snow came as high as his waist. Sometimes, he says, a family would kill their animals to put the beasts out of misery—and of course in winter it was easier to store meat. They would salt it and smoke it. No electricity, no refrigeration. None even today. Don Mario is still wary of slaughtering an animal for meat in summertime. He's more careful during those warm months; he eats more beans. And what about the lovely gardens surrounding his house? The rose bushes, the flowers, the trimmed shrubs and ornamental trees? It was Don Mario's wife, she was the gardener. She died two years ago from a cancer. Though he lives in the house alone now, he has family still in the valley, yes, two sons, grandchildren, an uncle. How much land? Don Mario is in possession of 377 hectares. But he fears that the lake—if there is to be a lake—will cover all of it. What then? If the lake comes, if his land and his house and the grave of his wife are covered over and he is driven away, he will go to Argentina. Not elsewhere in Chile, no, not this time. He will leave altogether, angry at his country and at the politics that could let such a thing happen. But is there a chance to prevent it? Yes, there is a chance, Don Mario believes. "It's luck that the gringos are here," he says in Spanish, with Spelius translating embarrassedly. "Because if it's not for that, the Chilenos would be on the ground, and stepped on." Looking me in the eye, Don Mario plants his foot and grinds it, crushing a piece of bark.

Will you win this fight? I ask.

"*Yo pienso que sí,*" he says.

WHAT MORE can I tell you about the Futaleufu River? I suppose I could describe those stupendous upper rapids, Zeta and Throne Room, which on a later day I view and appreciate from the safety of shore. I could mention the carcass of a forty-pound trout that lies stored—a tourist attraction now, a voucher for the local fishing—in a freezer at a little *hostelería* down the valley. I could recount the day we spend running the Rio Azul, a frolicsome stream in contrast to the demanding intensity of the big river. And I could end with another episode of my adventures and misadventures in Terminator. I could describe my final run through that rapid, on my last afternoon of paddling the Futaleufu—sneaking again down the left, pausing again in the crucial eddy, staring again at the horizon line, then launching myself over the ledge. I could tell you about the thrill of running Terminator right side up, about the rapture of threading down through its unholy chaos with the requisite strokes and braces and leans, about the satisfaction of reclaiming just a bit of my confidence and face. But somehow, in the context of other matters, that doesn't seem important.

[11]

# *The* CITY

# *The White Tigers of Cincinnati*

When I was a boy, four decades ago, the Cincinnati Zoo contained carnival rides. There was a low-amplitude and rather mild roller coaster with a canopy top like on a Chevy convertible, which sprang up to wrap the riders in darkness. It moved ripplingly around on its circuit, this thing, a huge slinky tube of metal and cloth, and my dim recollection is that it may have been called the Caterpillar. I'll never forget my first glimpse of it, but don't hold me to details, I was five. The Caterpillar loomed as a great wonder in five-year-old eyes, almost more thrilling to look at than to ride. I remember its astonishing wavy motion and its metallic roar; I remember the chatter of that wind-whipped canopy in duet with human squeals of glee. Eventually I was taken aboard it, no doubt, but what I mainly recall is that first distant view. The other zoo rides haven't stayed with me so vividly. A miniature train. A merry-go-round, possibly. Coney Island (the Ohio version, not the New York) was a full-blown amusement park that offered more intricate and excessive forms of vertigo, but Coney Island was many miles east of the city, down on the river, a once-per-summer treat, and to a child greedy for thrills the zoo was a fair substitute. Also, of course, the zoo contained animals. I liked animals as much as the next five-year-old, and maybe more.

Stolid rhinos and giraffes stood around on bare dirt. Lions paced away their nervous boredom behind bars and glass. Big snakes slept coiled like tugboat rope, but if I looked closely I could see them breathe. Elephants lived in an aromatic elephant house with a high domed roof, oddly resembling Monticello. Polar bears gasped from the midwestern July heat and swam in a pool of green water. They were, in their desolate extraction from context, a little greenish themselves. Sea lions performed on a wet stage at regular intervals, horn players without a string section, better musicians than I ever managed to become. Monkeys on an island. Crocodiles in a pit. There were tigers too, but in those days a tiger was black and orange.

It was an age of innocence. Tomatoes were juicy. Milk came in glass bottles. Tigers were colored to resemble the vertical patterns of a tropical woodland.

Times change. The Caterpillar is gone. Conservation and biodiversity are now watchwords at the Cincinnati Zoo.

Endangered species are housed and bred there. Many of the outdoor displays are cleverly designed to suggest pockets of habitat rather than fenced compounds. You can walk through a hothouse approximation of rainforest. You can gawk at an everloving abundance of live invertebrate animals—spiders and scorpions and cockroaches and butterflies—finally given their place in the limelight. In many ways it's an exemplary institution, this zoo: scientific, well-meaning, astute, and internationally recognized. Some things don't change, though, and the crass imperative of box-office appeal is one of them. In place of carnival rides, today, the Cincinnati Zoo contains white tigers.

White tigers are gorgeous creatures, and anyone who sets eyes on one will have a sense of beholding something extraordinary. Though that much is safely said, virtually everything else about them is controversial.

During the past twenty years, Cincinnati has been the birthplace of more white tigers than anywhere else in the world. It's no accident, and it's certainly no datum of biogeography. White tigers have been systematically bred at the Cincinnati Zoo, in roughly the same spirit as dachshunds are bred elsewhere. The white tiger program is highly successful within its own terms, but not universally applauded; professional zoo people have argued about whether it's a triumph of wild-animal husbandry or a travesty. "White tigers are freaks," according to William Conway, general director of the New York Zoological Society, a famously harsh judgment from one of America's preeminent zoo-based conservationists, who added: "It's not the role of a zoo to show two-headed calves and white tigers." Edward Maruska, director of the Cincinnati Zoo, objects to the term *freak* and prefers to consider them "off-tone genetic specimens" that occasionally turn up in the wild and therefore deserve also to be represented in zoos. Representing a certain mutational form, though, is not precisely the same as selectively breeding to proliferate that form.

Should these aberrant, handsome animals be propagated and protected in captivity? If so, why? Should it be done in the name of saving something natural and precious and wild—just as other zoos might propagate whooping cranes (*Grus americana*) or white rhinos (*Ceratotherium simum*) or the white starling (*Leucopsar rothschildi*) of Bali? Or are there other rationales, less lofty but more pragmatic? Do the lofty and the pragmatic rationales reinforce each other, or are they at cross-purposes?

Is white tiger propagation a legitimate enterprise in zookeeping? Or is it show business? Or is it perhaps both? The issue is complicated because those questions lead to others, some of which are easy to answer and some hard.

The easy questions are the scientific ones. For instance: What is a white tiger?

The hard questions involve social anthropology entangled with aesthetics and history and philosophy. For instance: What is a zoo?

THE WHITE tiger is not an endangered species. Nor is it even an endangered subspecies, like *Panthera leo persica*, the Asiatic lion. Some people confuse white tigers with Siberian tigers, possibly on the sensible though erroneous grounds that a subspecies of tiger native to snowy country ought to be white. But the Siberian tiger is yellowish in winter, orangish in summer, camouflage-relevant color shifts that are distinct from the phenomenon of whiteness. The Siberian tiger *is* an endangered subspecies, *Panthera tigris altaica*, native to northeastern Asia and larger-bodied than other tiger subspecies. A white tiger is something quite different: a mutant.

More precisely, a white tiger is an individual animal endowed with a double genetic dose of a particular mutant gene (call it a recessive allele) that causes (only in the double-dose situation) partial albinism. A white tiger has blue eyes, a pink nose, and creamy white fur with pale chocolate stripes, all symptomatic of its inherited deficiency in pigment. Not uncommonly, a white tiger may be cross-eyed. Strabismus, the experts call it. This congenital malady, in which the two eyes can't be aimed at a single point, is related to an abnormal arrangement of visual pathways in the brain. Both the pathway disorder and the cross-eyed condition are often associated with albinism—in Siamese cats, ferrets, mink, and various other mammals, including white tigers. White tigers (or at least the males) also tend to be larger in body size than most other tigers, an accidental similarity to the Siberian subspecies. A mutation for partial albinism could potentially occur among any population, but the white tigers that have become zoo and circus celebrities during this century are descended mainly from the Bengal subspecies, *Panthera tigris tigris.*

A Bengal tiger with two normal alleles at the given gene site is colored normally. A Bengal with one normal allele and one "whiteness" allele also looks normal. Only a double dose of the mutant allele (the genetic equivalent of rolling snake-eyes) results in manifest whiteness. Since a mutant allele at any particular gene site tends to be only sparsely distributed among a wild population of animals, and since close relatives share more of the same alleles (even the rare ones) than strangers do, the surest way to produce a double dose in any newborn is by inbreeding.

The grand patriarch of white tiger collections in modern zoos was an animal known as Mohan, captured as a cub in 1951 from a forest at Rewa, in central India. Mohan spent his life in a palace, as the coddled pet and genetic plaything of the last Maharajah of Rewa. Mohan's first offspring—fathered on an orange Bengal female—were as normally orange-colored as their mother. One of those cubs was Radha, a female who evidently carried (but didn't show) a single dose of the mutant gene. When Mohan was bred with

Radha, his daughter, she gave birth to four white cubs, the first generation of captive-born white tigers in this century. Among the four was another female, Mohini, so named to echo the name of her father, who was also her grandfather. *Mohan* is Sanskrit for "one who charms." It seemed apt to the Maharajah and it seems apt today. *Mohini* has been translated as "enchantress." True enough, there was something about these animals that would charm and enchant zookeepers and other folk into taking leave of their common sense. Mohini and her lineage became renowned as the White Tigers of Rewa. In due time she was bred to an orange-colored male, Samson, who was both her uncle and her half-brother. Being Mohan's son and Radha's brother, Samson seemed likely to carry a dose of the "whiteness" mutation. The pairing with Mohini revealed that he did, as together they produced another white cub, plus some orange-colored carriers. The Maharajah of Rewa by this time had sold Mohini to the National Zoological Park in Washington, D.C. The NZP loaned two of Mohini's offspring to the Cincinnati Zoo, where they were bred to each other—brother with sister—and produced white cubs. The sister, named Kesari, stayed on in Cincinnati long enough to become founding matriarch of the white tiger collection there.

In Cincinnati, the inbreeding continued. Bhim, a white son of Kesari, was mated successively to two of his sisters, Kamala and Sumita, the second of them especially notable in this genealogy. A white daughter of Kesari, Sumita has been described by Edward Maruska as "one of our prime breeding females," an understated tribute to an animal who has delivered twenty-five live-born white cubs—almost certainly more than any other female tiger, wild or captive, in the history of the world. If she lived at large in a tropical forest, a white tiger would probably never experience the peculiar circumstances that have led Sumita to her record. Tiger territoriality and the dispersal of young males tend to discourage inbreeding in the wild. Sumita, by contrast, has produced all of her white cubs from matings with Bhim, her brother.

Inbreeding has its costs, and strabismus is only one of them. Among the others are susceptibility to disease, lowered fertility, raised incidence of stillbirths, reproductive deformities, skeletal deformities, and loss of adaptability to new circumstances. Sometimes the negative effects can be mitigated—for instance, by intermittent outbreeding with another family line—and sometimes a severely inbred population is lucky enough to survive for decades without showing evidence of its genetic impoverishment. Outbreeding insofar as possible is a crucial part of most captive-propagation programs on behalf of endangered species. But white tiger programs, driven by other mandates, are run differently. These programs, including Cincinnati's, have tended to neglect outbreeding, because each outbreeding slows the rate of production of white cubs. Edward Maruska has argued emphatically that Cincinnati's white tigers have suffered little, at least so far, from the predictable problems of inbreeding. Possibly he's right. Possibly his program has been exceptionally skillful and exceptionally lucky. He can claim greater success at this peculiar task than any other zookeeper on the planet. But he can't change the laws of genetics.

What's the real purpose of inbreeding a family line to produce dozens of white tiger cubs? I carried that question back to the Cincinnati Zoo and knocked on an office door. I asked Bob Lotshaw, a curator of tigers, and he gave me an honest answer. "It's marketing. It's popularity. It's a major source of income for continuing other programs here at the zoo." White tigers boost zoo attendance in any city, but especially in Cincinnati, where even the professional football team is named after a subspecies of *Panthera tigris.* And in addition to gate receipts, there's the revenue from sale of excess animals. Cincinnati has sold its extra white tigers to zoos in Japan, South Korea, South Africa, Mexico, Canada, and Indonesia. It has also sold to Siegfried and Roy's Magic and Illusion Show of Las Vegas, Nevada. Slightly defensive on this point, Bob Lotshaw assured me that Siegfried and Roy give their tigers high-quality care and affectionate training, and I have no reason to disbelieve him.

The going price for a white tiger is $60,000.

ONE OF the sorry duties of a writer is to bite, every so often, the hand that has fed him. This is my relationship to the Cincinnati Zoo. I've returned there many times over the past forty years, and most of those enjoyable hours did not involve carnival rides. I've gone there repeatedly with people I love and I've gone there alone when I needed a refuge. It's an enclave of great beauty and apparent tranquillity. It wasn't the first place I experienced a sense of the wonders of nature, but it was certainly the first place I saw a bear or a tropical snake or a tiger. Maybe it played some sort of formative role in my life; or maybe not. Anyway, when I offer nasty cold judgments about this particular zoo, and about the dynamics of its appeal to the public, and for that matter about zoos generally, what I'm engaged in, partly, is an examination of conscience. Don't gag, we're not concerned here with an issue so small as my conscience, but I wanted to assure you that it's dirty.

Now to the larger question. What is a zoo?

How did these institutions come into being and why do we continue to tolerate them? What logic can be offered for keeping tigers—white ones, orange ones, tigers of any subspecies or color or genetic persuasion—on a small patch of grass behind a high fence in a sycamore-shaded neighborhood of Cincinnati?

Zoos are educational, people say. Zoos represent the only glimpse of nature that many city-bound children, and their city-bound parents, will ever get. Zoos are the last refuge of certain endangered species. Zoos are a way to preserve a few tiny fragments or vestiges of *wildness* itself, people say. The better zoos of today, including Cincinnati's, even play an important role in the captive propagation of rare and beleaguered animals that may eventually be released back into their native habitats, if by some miracle those habitats haven't meanwhile been totally destroyed. People say each of these things. I'd like to agree, I try to agree, but in the back of my brain and the pit of my stomach I don't. To me it's all half-truth.

The pedagogic value of zoos is an afterthought, dreamed up and added on during just the last 150 years to raise the intellectual and social tone of a much older tradition, the commercial menagerie. Even today, zoos aren't very educational. The white tiger exhibit in Cincinnati, for instance, is drastically coy on the subject of population genetics. Nor do zoos constitute fragments of wildness. In fact, wildness is precisely what's missing; the infinite intricacy of an ecosystem is missing; only the animate bodies of a few animals, some of those animals potentially dangerous to humans (which is not the same as being wild), but stripped of their contexts and their community roles and therefore their living identities, are present. Zoos do provide glimpses of biological exotica that can be taken to represent nature, it's true. But like many of the nature documentaries on public TV, zoos may actually undermine the continued existence of what they purport to celebrate. People watch the films, they visit the zoos, and by the mesmeric power of these vicarious experiences they come carelessly to believe that the Bengal tiger (or the white rhino, or the giant panda, or the diademed sifaka) is alive and well because they have seen it. Well I'm sorry but they *haven't* seen it. They've seen images; they've seen taxidermy on the hoof. And the wellness, even the aliveness, is too often a theatrical illusion. Zoos are not fragments of the world of nature, no. They are substitutes.

That's why they're useful and that's why they're pernicious.

Let's start again. What is a zoo? Most essentially, it's an arena of the visual. It's a place to see wonders. The act of seeing is the primary zoo experience—whereas learning, thinking, and emoting are dimensions of encounter that come secondarily, if at all. We go there to *look;* in passing, we read a few labels and placards, of which the information content is low. William Conway, having made that remark about "freaks," continued with a statement that frames zoos (even great conservationist zoos, such as his) in visual terms: "We need to save our severely limited space for tiger subspecies that are close to totally disappearing in nature. If we choose to save the white tiger at the expense, say, of the Siberian tiger, it's like saving a copy of a Rembrandt painted in glitter on velvet and throwing out 'The Man with the Golden Helmet.' " A fellow named John Berger would take the argument still further.

Berger is a cantankerous British art critic who has written about the viewing experience of zoos. He's well qualified to address the subject because a zoo, after all, as William Conway implied, is more like an art museum than like a forest.

What we see in a zoo, according to John Berger, are creatures that have been rendered marginal. "The animals, isolated from each other and without interaction between species, have become utterly dependent upon their keepers." So they are no longer wild. They are no longer complete. Their responses, their behavior, probably even their sensory capacities have changed. "Nothing surrounds them except their own lethargy or hyperactivity." Each has been separated, not just from its natural habitat, but from its identity. They are numbed; in some sense, as they stand or sleep or eat or pace the cage floor, they are already extinguished. To assume that they retain the capacity for seeing and experiencing us, while we are seeing and experiencing them, is recklessly hopeful. Some writers have argued that the value of a zoo visit comes when a human and another animal make contact with their eyes. Forget about that, says Berger, it doesn't happen. "At the most, the animal's gaze flickers and passes on. They look sideways. They look blindly beyond. They scan mechanically." They have been immunized against encounter, he says. When a human looks deep into the eyes of a zoo animal, according to John Berger, the human is alone.

I wasn't so sure. I wanted to look deep into the eyes of a white tiger. So, after talking with Bob Lotshaw, I walked back up to the compound in which four of the Cincinnati animals were basking beneath a wan winter sun.

They were confined to about a third of an acre of grassy slope, with a few rocks and bare trees—no prey, no competitors, no territorial structure, no forest, no natural context—but they were making the best of it. Occasionally one would pounce on another playfully. Their noses were pink. Their fur was creamy with umber stripes. They were impossibly, pathetically beautiful. If their eyes were blue, or otherwise remarkable, they didn't show me.

The fence was eighteen feet high. Above was a wooden walkway, from which I and the few other visitors could look down. It was a quiet day in February—a day of low attendance, despite these box-office stars. Traffic sounds came from a distance, muffled by hills and trees. Suddenly a young boy on the walkway started screaming. The forced anger of his shrieks indicated a tantrum. He wanted his lunch, or an ice cream, or a Coke, or his nap, or simply his way on some other point. Maybe he wanted a carnival ride. He was about five. His squalling caught the attention of two tigers, below, and I watched as they stalked him along the fence. Their heads went low and they stepped carefully. They seemed to have come alive. Part of me hoped to see them leap up, clamber over the fence in one astonishing burst of athleticism, snatch that kid, drag him back down into their compound, and eat the little snot. But I was divided.

Then the tigers remembered their own hopeless impotence, and they lost interest.

# *To Live and Die in L.A.*

The cab driver's name is Amir, he's a young Israeli, he lives in Sherman Oaks, just over the so-called Santa Monica Mountains from Beverly Hills. He meets me at my hotel, on Wilshire Boulevard, and for a moment we confer over destination and routing. Yes yes, says Amir, he believes he can find Burbank, no problem. He is a professional, with a professional map. So we drive off along the rain-slick Los Angeles freeways. At 5:30 on this winter morning it's still dark, but the roads are already astream with traffic, and I'm happy to let Amir prove his professionalism versus the ramps and the merging lanes and the welter of split-second choices while I concentrate on my coffee. Thank heaven for taxis. If I were driving, I'd commit that single wrong move and wind up gridlocked on the far side of Long Beach at midmorning, whereas it's crucial to reach Wildwood Canyon Park before dawn.

"You are doing what?" says Amir.

Going to look for coyotes, I tell him.

There are no coyotes in Tel Aviv, just one thing among many that distinguish Amir's home city from Los Angeles. There are no coyotes in Israel whatsoever; the world's most cunning and adaptable species of canid is strictly a creature of the Americas. But Amir is a quick study, and when I repeat the word, it registers. Kee-yotee, he says. Ah, like a dog, yes? That's the one, I say: Like a dog, only it's a wild animal. Amir is amused. He appreciates this departure from cab-driverly routine. Most of his early-morning fares are not gentlemen headed out to slog in the hills above Burbank in search of coyotes, he informs me. A silent ride to the airport is rather more common. While Amir talks, I gape out the window toward Studio City, noticing that the drizzle has gotten heavier. Then we make our appointed crossover from the Ventura Freeway to the Golden State Freeway, swinging back toward the northwest, and Amir adds: "I don't know, these people, in this city, if they know that they *have* coyotes."

No doubt he's more right than wrong. Coyotes are adept, when they want to be, at remaining invisible. Within the Los Angeles basin, they're obliged to choose their home territories and their travel corridors with care, finding security in the large patches and linear strips of greenery preserved as parks and wisps of landscaping, or in the topographic anomalies too steep and too fashionable to be buried completely by pavement: patches like Griffith Park, linear strips like Mulholland Drive, topographic anomalies like Laurel Canyon. These remnants of habitat are fragmented and scattered, but they aren't scattered uniformly throughout the city. A person could live a lifetime within certain neighborhoods—Inglewood or Chinatown, for instance, or Mar Vista or Watts—without catching sight of a coyote. But while most of the Los Angeles populace may not know that they're sharing their metropolitan area with *Canis latrans,* some people certainly do. Encounters happen. The city's Department of Animal Regulation can bear witness to that.

In the course of an average year, the Department of Animal Regulation takes about seven hundred complaint calls involving coyotes. *A coyote ate my cat,* someone says. *A coyote murdered my cockatoo, Orson, while he was sunning on the patio,* whines somebody else. Or: *A coyote jumped the back fence and pushed over a garbage can and scarfed up three grapefruit rinds and a quart of cappuccino yogurt, then glanced at my baby daughter and licked its chops.* Or maybe: *A whole pack of coyotes came swaggering down Sunset and scared the bejesus out of me and my Airedale.*

Or my own personal favorite: *A coyote charged in through the pet door and disemboweled our poodle on the kitchen floor.* Upon request, the DAR traps and destroys coyotes in any vicinity where a single animal has caused fear or trouble. Last year that amounted to roughly 160 dead coyotes within the city proper. The trapping procedures are not focused enough to target specific individuals, so there's no way of knowing how many of the 160 dead animals were executed for offenses they didn't commit, but a plausible case can be made, I suppose, that some minimal level of reactive control is within reason. Anyway, the persecution of coyotes in Los Angeles is not nearly so egregious as the persecution of coyotes in, say, the state of Montana. And those L.A. coyotes that remain innocent of garbage scarfing and poodle disembowelment, plus the ones that are just too smart to be trapped for their sins, can continue to thrive within the slightly less fashionable patches of urban greenery, such as Wildwood Canyon Park.

Amir drops me off at the park gate, with a promise to return in four hours. The gate is barricaded to traffic. A sign says PARK CLOSED, adding in large letters that violators will be SUBJECT TO ARREST OR CITATION, so I glance around carefully before ignoring it. If a car comes, I decide, I'll dive into the brush. The minions of civil order will never catch me. I feel alert and superior and mildly outlawish this morning, besides which I've got business to do and can't be bothered with arbitrary boundaries. In other words: One step past the barricade and I've entered the mind-set as well as the habitat of the Los Angeles coyote.

THE PARK road winds upward into steep hills and chaparral vegetation, away from the lights of somnolent Burbank. At the end of the road is a trail. At the end of the trail, on a ridgetop, I begin bushwhacking across to the next canyon. The earth of these slopes is soft and unstable, a porridge of mud-thickened sand that tends to slough away under each footstep. On a patch of bare mud I notice some tracks, unmistakably canid pad-and-toe prints but with no claw marks punctuating the outer two toes, a telling absence that suggests coyote rather than dog. I see a scat pile, each morsel pointy and fibrous at one end, wastage from what must have been a fur-covered little meal. This much seems promising.

So after topping over into the next canyon—which turns out to be roadless, pathless, and surprisingly unscarred by human intrusion, a tiny token of wilderness surrounded by urban sprawl—I settle myself down on a mat of dead grass. From here my view is commanding: steep slopes, a brush-choked stream channel below, another high ridge across the way, a network of game trails lightly etched through the chaparral and, visible just a mile to the west, beyond where the canyon opens out onto the flat, Burbank.

I raise the hood of my jacket, pull my hands back inside the sleeves, and hunker, reasonably cozy despite the chilly rain. I sweep the far slope with my binoculars. Nothing. But it seems a good place to keep watch, and an excellent place to sit contemplating the persistence of coyotes in Los Angeles.

They have inhabited this valley since long before *Homo sapiens* arrived, of course. That's a claim that can't be made everywhere throughout the present distributional range of *C. latrans,* which was largely confined to the prairies and deserts of western North America at the time Europeans first invaded. Nowadays there are coyote populations to be found from California to Maine, from Costa Rica up through Alaska; the species has spread eastward and northward and southward to the limits of landmass and tolerable climate, expanding its range in association with those human settlers who accommodatingly cleared forests, filled grasslands with tasty and bone-stupid livestock, and exterminated the coyote's two chief competitors, *Canis lupus* and *Canis rufus,* the gray wolf and the red wolf. But in the Los Angeles basin the coyote is no newcomer.

Fossils taken from the Rancho La Brea tar pit (one of the great sites in urban paleontology, back down on Wilshire Boulevard not far from my hotel) suggest that *C. latrans* was present here during the Pleistocene, sharing its habitat with dire wolves and sabertooth cats. The dire wolves in particular, gigantic and now-extinct predators of the species *Canis dirus,* serve as a reminder that maximal size and ferocity are not always favored by the pressures of evolution. The sabertooth cats and the dire wolves are long gone, but the coyote was a more versatile animal—smaller in size, less specialized in anatomy and habits, smart enough for social behavior, with an opportunistic disposition and a high rate of reproduction, all of which helped it survive.

For the Serrano Indians, in the early Los Angeles basin, the coyote held an important totemic status, as it did for other tribes in the Southwest. Among the Serrano, the coyote gave a name, an image, a social identity to the *Wahilyam* ceremonial society. It seems to have been an ambivalent totem, though, representing some good aspects (resourcefulness, for instance) and some bad aspects (villainous trickery) of human character. During the Spanish mission period, the coyote population in southern California increased dramatically, not just because the Franciscan missionaries were raising big herds of cattle, horses, and sheep, but because the cattle were grown for hides and for tallow, not for meat. When cows and steers were slaughtered, great piles of beef and offal were left to stink up the landscape. Since a female coyote is capable of producing up to a dozen pups annually, given abundant food and adequate security, *C. latrans* was perfectly suited to proliferate while the excess of carrion lasted. (The California condor probably thrived during that period too, but as a slow breeder it couldn't increase its population so quickly.) The pattern of cattle raising, meat wasting, and coyote feeding continued after the missions were secularized, until a severe drought in the 1860s hit the tallow-and-hides industry hard enough to start a shift toward more varied agriculture. Within the Los Angeles basin, chaparral and sage were cleared for the planting of orchards, grainfields, vineyards. For coyotes, the great pig-out was over, but the new situation wasn't disastrous. Though they belong to the order Carnivora, coyotes are actually omnivores, well able to scrounge and survive on a wide range of food items including mice, ground squirrels, rabbits, snakes, insects, grapes, peaches, avocados, prickly pear fruit, peanuts, cantaloupe, watermelons, dates, celery, soybean meal, harness straps, and oranges, as well as poodles and Persian cats and the occasional wing-clipped flamingo. So they made the transition to modern Los Angeles easily.

A geographer named Don Gill studied the coyotes of L.A. during the late 1960s and early '70s. He estimated that at least four hundred animals were living inside the city, with several thousand more in the mountains along the immediate periphery of the basin. The Department of Animal Regulation, in those days, was hearing just over a hundred complaints annually (less than one-sixth as many as now) and killing thirty-five coyotes (versus 160 now) in an average year. That numerical increase during the past twenty years could reflect two things: a sheer increase in the population of coyotes or an increase in the chance that any given coyote will suffer a fateful encounter with humans. Either explanation might be true, and possibly both.

The animals, as charted by Gill, were clustered wherever brush and topography offered adequate refuge. About twenty lived within Griffith Park. There were scattered sightings and reported predations in Whittier, in Arcadia, in San Marino, and in Pasadena just west of the Rose Bowl. East Los Angeles, unaccountably, was a minor hotspot. Concerts at the Hollywood Bowl, an open-air theater surrounded with chaparral, were occasionally interrupted by howling. The Santa Monica range, a wedge-shaped zone of wooded hills jutting eastward into Los Angeles from the coast, with Mulholland Drive marking the ridge line and those fashionable canyons (Franklin Canyon, Coldwater Canyon, Laurel Canyon, among others) draining southward, harbored a big share of the city's coyotes. One other island of occupied habitat, according to Gill, was the Verdugo Mountains area, bounded by the Foothill and the Golden State freeways, just northeast of Burbank. Don Gill himself had found coyote spoor in the Verdugos.

The Verdugo Mountains encompass Wildwood Canyon Park. Almost thirty years have passed since Don Gill collected his data, with hillside dream houses and cantilevered patios usurping canyon habitat all over L.A., but Wildwood remains (in its modest and insular way) wild. Today I've got the park to myself. I sit in the rain, waiting and hoping for some evidence that *C. latrans* is still around. I wonder: How much longer will Los Angeles tolerate the presence of coyotes, and vice versa?

ON ANOTHER winter day, I sit on another steep hillside, watching a pack of four decidedly nonurban coyotes. They are prowling across a snow-crusted sweep of bottomland along the upper Lamar River in Yellowstone Park. These are handsome, grayish cream animals in good condition, contentedly absorbed in hunting for voles, small rodents that move through snow tunnels under the crust. The coyotes are oblivious to me, and to Dr. Robert Crabtree, the coyote researcher who has brought me here. This group of animals, according to Crabtree's nomenclature, is the Fossil Forest Pack. The female wears a radio collar. One of the younger males seems to be limping. To Crabtree, they're familiar individuals with a known family history.

Bob Crabtree has spent eight years studying the demographics and social structure of coyotes. He did much of that work on an ecological reserve in the sagebrush of southcentral Washington state; more recently, his site is Yellowstone. The coyote populations that he has sedulously watched and measured differ in one crucial way from most other coyote populations in the United States: Crabtree's have been "unexploited" populations, protected from hunting and trapping.

The institutionalized persecution of coyotes in this country over the past sixty years is a huge and outrageous story that deserves its own telling, but I won't launch into that here. I'll just mention that your federal tax dollars support a program called Animal Damage Control, administered within the Department of Agriculture, the main service of which is to kill about 76,000 coyotes per year. The beneficiaries of that service are ranchers, chiefly sheep ranchers, who graze their animals on both public and private lands. In addition to the ADC harvest, another 350,000 or so coyotes are killed privately by hunters and trappers in a typical year. The net result of all this coyote slaughter—Bob Crabtree suspects, and expresses in judicious scientific terms as we sit talking—is exactly the opposite of what's intended.

Destroy coyotes indiscriminately, he explains, and you destroy nonreproductive adults. You kill mature alpha males and alpha females (that is, pack leaders) in the six-to-twelve-year-old class, aging animals that are robust enough to maintain their dominant status and their territories but too old to be fertile. (In other words, these relatively few individuals hold the exclusive social prerogative to breed, but in many cases no longer exercise it.) Kill them and you encourage younger, submissive animals to claim dominance and begin breeding. You increase the reproductive rate of the population. Do that and you increase the predation against sheep. Why? Because young coyotes with newborn pups account for most of the lamb killing (or the killing of elk calves in Yellowstone), according to Crabtree. The growing pups need such windfalls of protein. Older coyotes without pups tend to take the easier, safer course, feeding on rodents, rabbits, and other reliable small-item foods.

Among the unexploited populations he has studied, Crabtree has seen the result of allowing those mature alphas to survive. "It's a very important form of population limitation," he says. "Because these old alpha females are keeping whole territories unproductive. And the consequence of that is lower predation on lambs."

The Fossil Forest Pack work their way downstream on the far bank of the Lamar, passing below our overlook. Through my binoculars, I watch the female pounce. She digs, she roots, she brings up a vole and swallows it. Several minutes later, she and her mate and the two younger males tilt back their heads and break into a chorus of yipping howls: voicing a territorial claim, no doubt, but also perhaps venting pure pride and satisfaction in their coyotehood, on a fine piece of landscape where they have never set eyes on a sheep.

Crabtree makes one further intriguing point. Sixty years' worth of largely indiscriminate killing by ADC trappers and others has inevitably reshaped the American coyote, producing an animal that's more clever and wary and resourceful, more problematic, than ever before. The fittest have survived, and doggone if the fittest aren't harder to trap, harder to poison, harder to fence out, harder to fool, harder to kill despite all the helicopters and leg-hold traps and high-powered rifles and cyanide booby traps that ADC can muster. "They've created their own worst nightmare," says Crabtree, not without sympathy for the many trappers and ranchers he has gotten to know. "They've created a coyote that's impervious to their means."

I SIT IN the rain, wondering: What sort of coyote has Los Angeles created?

It's a creature that will jump over chainlink for a bowl of Alpo. It's an animal that can learn and remember which storm-sewer channels lead to which golf courses, which duck ponds and swimming pools offer potable water when the hills are dry, which dumpsters behind which supermarkets are likely to be overflowing with old vegetables and delightfully rancid fish. It's a beast constantly on the alert for unattended barbecued chicken. It's a predator that, like some two-legged ones, is at home on Mulholland Drive. It has eaten from the Tree of Forbidden Knowledge, and it recalls fondly the taste of Fifi and Mr. Boots.

I confess that I find this neither sad nor inappropriate. Better for the people of Los Angeles to share their city with one slightly corrupted species of Carnivora, I think, than with none at all, and the coyote is ideal for the role. It's arguably more similar to *Homo sapiens,* in ecological terms if not anatomical ones, than any other species of animal, including the chimpanzee. And persecution by humans just makes it more similar still.

Will coyotes still haunt this city in a decade or two? I asked that question of a veteran trapper for the Los Angeles DAR. "Oh, definitely," he told me. "Twenty years from now? I'll guarantee it. A *hundred* years from now. They'll be around. They're survivors."

But after an hour's soggy vigil, I've spotted no coyotes on the game trails of Wildwood Canyon Park. Not this particular morning. And my hands have gone numb. Time to move. I descend to the stream bottom and start hiking downstream, shin-deep in the water, tunneling my way out beneath a fallen archway of riparian brush. I've got no other choice but to walk the stream, since the game trails are rambling and evanescent, and the chaparral-thick hillsides are impassible. On a low grassy bank, sheltered by an oak tree, I stop. I notice something there, partly buried in silt. It's a coyote skull.

Experts will later confirm that identification, though I'm fairly confident from the first glance. It's muddy but meatless, a graceful long-snouted Yorick nicely bleached by time and weather. I examine its eye sockets, its remaining few teeth. What's your story, skull? Maybe the animal it belonged to lived a long tranquil lifetime right here within the fastness of the Verdugo Mountains, never venturing down into Burbank, never robbing a barbecue, never killing a pet, never digging for moles on a fairway, never committing a single act of subversion or trespass against the human-made world. But naw, probably not. Probably this was another urban guerrilla. Still, the fact that its mortal remains came to repose here, on a stream bank in a pathless canyon, suggests that the critter was lucky or cunning enough to die a natural death.

Good. Feeling outlawish, willfully pilfering a natural artifact from Wildwood Canyon Park, I wrap the skull gingerly and put it into my bag. Then I slog on. They'll never catch me. I'm only a half hour from my taxi.

# *Reaction Wood*

I'll start with the memory part, leaving the science part for later. When I was a boy I had, in lieu of a dog, in lieu of a grandfather, a tree.

It was a towering old black walnut that stood near the south edge of our yard in suburban Cincinnati, with a pair of stout lower limbs sticking out horizontally like the arms on a giant scarecrow. If my memory can be trusted, those limbs emerged from the trunk about eight feet up. Possibly they were lower and only the parallax of time and nostalgia has raised them. Anyway, one held a swing. Supporting my weight, or mine and my big sister's together, would have been trifling to a limb of such girth—it was thick as a telephone pole. In the fog of my earliest recollections, dating back forty-some years, I can see myself gazing up at those two horizontal limbs, where they hung far beyond my little-boy reach, far beyond even my best jump. I can recall vowing that someday I'd manage to lay hold of one, somehow, and then I'd climb this wonderful tree.

Reaching the horizontal limbs, the first rung, would be the least of the problems. Just above, the tree divided into three major stalks, each nearly vertical, the easternmost of which rose to a precarious crow's nest of small branches about sixty feet in the air. Getting to the crow's nest would present dangers and difficulties whose solution, when I was six or seven, I couldn't even imagine. But several years later, with a little more size and agility, I solved them. And then, over and over again throughout boyhood, I did climb the tree. I went everywhere in it that my weight would allow. I learned all its knobs and its crotches. I discovered that although the crow's nest seemed unattainably high on its limbless stalk, I could reach it indirectly: ascending the westernmost stalk instead, then stepping over to a high notch in the middle stalk, and from there to the eastern one. I wore a path, along that route, into the tree's black corduroy bark. Once, on a stupid show-offish whim of the sort that occasionally gets a boy killed, I climbed to the crow's nest and down again blindfolded. I had memorized every requisite move. But the tree represented more to me than a gymnastic challenge. It was a place I enjoyed visiting; it was a living creature I respected. It had become almost personified to me, a valued companion and mentor.

Probably it was then about a hundred years old. That would have made it a sapling in 1864, when Ulysses S. Grant and William T. Sherman met in a Cincinnati hotel room to sketch out the last campaign of the Civil War. By the time my family took custody of the tree, it was no longer subsumed within a continuous hardwood forest rolling over the hills at the near northern fringe of the city. It had been spared when the forest was otherwise cleared, left to loom lonesomely over a farm landscape and then, later, over a half-acre lot. When my parents built a house on that lot in 1950, all that remained of its ecological and agricultural vestiges were the walnut tree, a hidden well shaft lurking treacherously beneath a layer of rotten timbers and soil in the backyard, and an 1825 penny, which my father found in the dirt. But the hardwood forest—or a large remnant of it, at least—survived nearby.

Beginning just beyond a thistle field that began just beyond our back fence, the forest stretched for miles, a boy's wonderland of trees, vines, hillsides, trails, and rocky creeks full of crawdads and frogs. The northern borderlands of the city also included a local park known as Winton Woods and another called Mount Airy Forest, two parcels of protected landscape where my father and I occasionally hiked, skied, or fished. But those two were relatively distant, reserved for special excursions by car. The woods just behind our house was different in several ways: more accessible to me, more wild, a better-kept secret, unofficial, unprotected. I never did know who owned this suburban wilderness but, as events would prove, whoever the owners were, to them it was just real estate temporarily encumbered with trees.

Also unlike the two parks, it was nameless. I never called it anything but "the woods." Where have you been? my mother would ask. The woods. What's in the bucket? Um, salamanders. Where did you get them? The woods. Who's in the box? Um, a turtle. What's that inside your shirt? Um, a snake. This wasn't disapproving interrogation; she knew she was raising a zookeeper, an explorer of mud wallows, a feral primate, and her indulgence was heroic. What's your plan for Saturday? I dunno, I'd say, I guess maybe I'll go to the woods. In truth, I spent so much time there that eventually I needed remedial training in basketball.

MEANWHILE, closer to home, the walnut tree and I got to know each other better and better. Its fruit fell as heavy green globes, hefty as golf balls, excellent for throwing at garbage cans. Each globe was wrapped in a thick husk that could stain your hands nicotine brown with its juices. The smell was unforgettable—turpentiney but fresh. Inside the husk was a round nut with a rugose surface, hard as brick, and inside the nut was meat. The tree's leaves were compound, with multiple leaflets attached in opposing rows down both sides of a wand-thin stem. In autumn my chores included raking those leaves and stems and gathering those nuts so that the lawn mower, on its last pass of the season, wouldn't launch them like shrapnel. One autumn my parents and sisters and I made a project of harvesting them. The nuts were left to season in grocery bags until their husks dried and darkened. Then we cracked them with hammers and teased out the meat. Many hours' work, wrapped in that fine walnutty aroma, yielded a small pile of meat fragments and an entire suburban family with hands dyed brown. Besides these various ministrations to the tree, I continued climbing it.

At one point, having outgrown the little-kids' swing, I hung a trapeze from a higher limb. But I never built a tree house or defaced the trunk with a ladder-line of plank steps. Driving nails into this living xylem would have seemed barbarous, both to me and to my father, himself a great lover of trees. The tree in its unsullied state embodied all the stairway and structure that seemed necessary. Besides, a board shack furnished with old carpet and comic books and a flashlight, forty feet off the ground, was not what I wanted. What I wanted was a tree. When the urge struck me to daydream, or to pout, or to gaze out across the landscape, I went to the crow's nest.

And there was a further attraction: I was an acrophile. I liked the scary thrill of attaining heights. Though I didn't know it then, the walnut tree was my tiny midwestern substitute for granite spires and walls in the Sierra.

The tree's lower reaches, though less thrilling and less solitary, weren't neglected. As my balance became steadier, I took idle pleasure in tightroping on the two horizontal limbs. One especially, the tree's right arm as facing the yard, longer and straighter and more bare than the other, made a nice sort of balance beam. I would walk out the length of it, some thirty feet, then hang and drop from a branch there, or else turn carefully and tiptoe back to the trunk. In summer I'd do it barefoot, getting a better touch of skin against bark.

At the time when this stunt amused me, I was no longer such a little boy. I must have weighed at least 110 pounds. I didn't realize until recently what that extra load might have meant to the tree.

TREES ARE big creatures that live a long time, supporting vast weights of themselves at various splayed angles against the steady tug of gravity, the occasional burden of ice or snow, and the intermittent shoving and twisting of wind. As they compete with one another for sunlight, water, soil nutrients, and space, that competitive struggle goads them upward and outward. But craning upward and outward in midair isn't easy. Beyond the basic ecological difficulties of survival, trees also face severe mechanical stresses—stresses that a jellyfish or a strand of kelp or a giant squid, living within the supportive medium of seawater, is never forced to endure. For a tree, standing high and dry, life isn't so restful as it might seem. Great strength, supple resistance, and prolonged exertion are required. Imagine a chin-up that lasts a century.

A sizable body of scientific literature has developed in recent decades on the biomechanical design of trees. These studies concern patterns of limb and trunk structure and how those patterns reflect the imperatives of growth and survival. There's a classic little book called *The Adaptive Geometry of Trees*, full of calculus and theory, published twenty-five years ago by a very fresh-minded biologist named Henry S. Horn. There are several interesting papers by Thomas A. McMahon, a professor of applied mechanics at Harvard. There's an influential volume titled *Tropical Trees and Forests: An Architectural Analysis*, by Francis Hallé and two other scientists. As epigraph to their book, Hallé and his coauthors offer a quote from the venerable botanist E. J. H. Corner: "Botany needs help from the tropics; its big plants will engender big thinking." The implied point is that *Tropical Trees* is not just about tropical trees but about the architectural principles common to all trees, as revealed most vividly in the tropics.

Why is the shape of a mangrove so different from the shape of a ceiba? Why is a baobab so different from a pandanus? Crossing back up into the north temperate zone, why is a pandanus so different from an elm or a maple? Why is a maple so different from a fir? Why is a fir different—noticeably, if not drastically—from a pine? The variety of shapes, and therefore the number of questions in this vein, might seem almost infinite. But the variety *isn't* infinite, and some scientists have reduced its myriad particularities to a tractable number of generalized patterns. One authoritative source, *Trees: Structure and Function*, published a generation ago by two professors of forestry named Zimmermann and Brown, describes three basic tree forms: columnar, excurrent, and decurrent.

A columnar tree is one that rises in a single bare trunk, a column, topped off by a tuft of leaves. For instance, a coconut palm. Life is a little simpler and more limited in scope for columnar trees, due to the absence of limbs. An excurrent tree is one with a central stalk that predominates over a number of lateral branches, resulting in a cone-shaped profile. Most conifers are excurrent. A balsam fir carefully cultured and sheared for the Christmas-tree market embodies the very ideal of excurrence, its central stalk tapering up into an apical shoot to receive the terminal star or the glass angel. A decurrent tree shows a diffuse pattern of branching in which no central stalk remains dominant; instead, the trunk splits into a number of roughly coequal limbs fanned out through a full 180-degree arc, giving the crown a rounded profile. Maples, elms, hickories, oaks, and most other hardwoods are decurrent. The black walnut, *Juglans nigra,* is decurrent. If it weren't, the crow's nest I climbed to wouldn't have been so far off-center, and the horizontal limb I walked back and forth on wouldn't have been so stout.

Francis Hallé and his collaborators went beyond this three-category system in the 1970s, when they proposed their concept of "architectural model" among trees. What they meant by that term was "an abstraction of the genetic ground plan upon which the construction of the tree is based." They defined twenty-three different models, enough to encompass all tropical and temperate trees, naming each model after a botanist who had contributed to the understanding of species within that particular group—Corner's model, Rauh's model, Scarrone's model, and so on. The distinguishing characters included such factors as whether the trunk puts out any branches whatsoever, whether primary branches split further into secondary branches, whether new shoots grow continuously or in cycles, where flowers occur, and whether branches angle themselves more or less vertically or horizontally. All these factors relate directly to the growth process, and not quite so directly to the tree's eventual shape. A tree's architectural model, according to Hallé and company, is merely its genetically programmed "plan of growth." The actual structure of an individual tree at its peak of maturity, or in its dotage, would reflect not just the genetic plan but also the accidents, the environmental circumstances, and the stresses of a lifetime. One bolt of lightning, or a typhoon, or an overfed orangutan putting too much weight on a branch could cause a permanent morphological modification. Traumatic injuries might heal, but evidence of those injuries would remain recorded in the tree's physical structure. Zimmermann and Brown, with their simpler schema, had made the same fundamental point: that the shape of a given tree represents an interaction between destiny and experience.

One aspect of that interaction is the maintenance of limb position. Gravity tends to pull limbs down, especially as they grow thicker and heavier. The tree's quest for sunshine, implemented by growth hormones that bend shoots against gravity and toward light, tends to angle them up. Another hormone-moderated process, known as epinasty, tends to push limbs away from the tree's vertical axis and therefore again also down. The adaptive value of this epinasty business seems to be that it keeps limbs from crowding one another, so that the tree won't waste resources by competing with itself for space and light. Somewhere between these opposing forces, the limbs come to equilibrium. Whether the tree is excurrent or decurrent in form, whether it reflects one architectural model or another, each limb has its preferred position and tends to maintain itself there.

If a new external stress is applied against a limb, tugging it downward or crimping it upward, the tree responds. It marshals resistance. It produces what botanists call reaction wood.

REACTION wood is a muscle-like extra thickening that arises in the radial dimension of growth, exerting an opposing force against any distortion of the tree's natural shape or posture. It tends to raise limbs, or lower them, back toward their normal positions. It can even be mustered to support an entire trunk that's been tilted off vertical and to bend at least the more flexible upper end of that trunk back upright. Its existence has been well established by experiment. Botanists have tied branches akilter—bending them upward, bending them downward, bending them around into full loops—and then, after several seasons, examined the growth patterns in cross section. They have found that reaction wood begins accumulating promptly as the tree struggles to correct its disarrangement.

How does reaction wood work? There's no single answer. An odd thing about this phenomenon is that the same result is achieved two different ways by two different types of tree. Conifers produce one sort of reaction wood, known as compression wood. Hardwoods produce another sort, known as tension wood.

Compression wood appears on the underside of a limb that

has been displaced downward, tending to *push* it back up. Tension wood appears on the upper side of a limb that has been displaced downward, tending to *pull* it back up. Likewise if the limb is displaced upward: Compression wood in a conifer will tend to push it—tension wood in a hardwood will tend to pull it—in each case, back down. Compression wood is more dense than normal wood, with thicker cell walls, a higher content of lignin, and other differences at the cellular level that generate expansive growth strains. It's also distinguishable by its reddish color, at least when freshly cut. Tension wood is less dense and less lignified, with gelatinous fibers in the cell walls and other differences that generate contractile growth strains. It isn't so easily recognized by color, but it can be seen clearly in a cross-sectional slab of asymmetrical growth rings, like a bulging biceps on the upper side of an arm. Both types of reaction wood are promoted by the same group of growth hormones, known as the auxins, though there's still some uncertainty as to how the auxin works in each case.

Why this basic difference exists—conifers reliant on pushing, hardwoods on pulling—is another mystery that botanists don't yet seem to have solved. Personally, despite my curiosity, I'm glad that the trees are keeping some secrets.

IN A BLACK walnut tree, then, reaction wood will form on the upper side of a horizontal limb subjected to extra downward stress. Over time, assuming that the stress doesn't cause a crack or an outright collapse, the net result might be a stronger limb. Having developed an additional thickness of resistant tissue, it might be better able to withstand the routine vicissitudes and episodic crises of later life. But no, I'm not going to try to persuade you that my weight on the limb of the old tree in Cincinnati had that effect.

The tree passed out of my family's custody in 1966. I've gone back to peek at it several times, most recently about a decade ago. Last week, feeling curious again, I made a call to get news. Yes, the old tree is still standing, my father reported back; it looks pretty healthy, though some of the limbs have been pruned. Which limbs? I asked anxiously. Well, the big horizontal one on the right is gone totally, he told me.

So much for my contribution to the tree.

I've been thinking more about its contribution to me. It amused me, it nurtured me, it challenged me, it taught me in ways I can guess at but not measure. It showed me the possibility of a deep fondness for an individual living creature beyond the usual channels of sentiment—that is, beyond the easy reciprocal relations of boy-to-dog, girl-to-horse, boy-to-snake. It lifted me up, many times. And from the vantage point of its crow's nest, I saw for the first time but not the last a forest destroyed.

I saw bulldozers come, tear a route just past the base of that walnut tree and out through the thistle field and on into the woods. I saw trees felled wholesale and scraped away to make space for humans. I watched the woods disappear, becoming suburb. This happened for all the usual, unimpeachable reasons: People had money to spend, people had new mobility, people had growing families, people wanted nice houses with little yards in front. Roughly the same reasons had led my own parents to the half-acre lot with the walnut tree.

It's complicated, I admit. We all carry our shares of responsibility, large or small, obvious or indirect, for the loss of the planet's wild places. We burn fuel, we eat food, we consume extravagant quantities of other resources whose production and delivery take a fateful toll, and most consequentially, we bear children whose needs and demands will vastly multiply our own impacts through time. Who's to say what is or isn't a "reasonable" amount of personal self-interest or an "acceptable" amount of further attrition, justified on the grounds that, say, jobs are at stake? Who's to say that this or that forest shouldn't be cleared for a solid, pragmatic purpose suiting one group of people or another? That this or that river shouldn't be dammed or dewatered or channelized? That this or that species of butterfly or lizard shouldn't be consigned to extinction? That this or that compromise shouldn't be made? We all need to be willing and able to bend. Don't we?

But at some point, watching from high in a walnut tree or elsewhere, a person may stiffen. Tissues of resistance may arise. It's a human sort of reaction too: Enough, I bend this far, no farther.

# *Superdove on 46th Street*

As I step out the door of the hotel and turn east along the canyon of 46th Street, toward the shadowy heart of Manhattan, I see a pigeon. Needless to say, this would not usually constitute a noteworthy ornithological observation. But today isn't usual.

The bird is ordinary enough in appearance, and it's behaving as city pigeons do. It shuffles sideways with no sign of nervousness, letting me pass. It knows the rhythms and protocols of New York street traffic better than I do, and it's less of a stranger in a strange land. Its head ticks forward and back as it moves, counterbalancing the motion of its legs. By any unprejudiced standard it's a handsome animal, feathered plushly in Confederate gray, with a pair of blue bands across each wing and a sheen of green iridescence on the nape of its neck—the archetypal pigeony color scheme. Still, I see something different in this creature from what I saw the last time I looked at a New York City pigeon. The bird itself isn't aberrant but my perspective has changed.

"There it is," I think. "The superdove."

I feel a mild sense of awe, and a tinge of eschatological concern. It doesn't look like a creature of such menacing superiority;

it doesn't look like a conqueror of worlds. But lately I have cause to wonder.

THE "SUPERDOVE" business comes from a new book I've been reading, by two biologists named Richard F. Johnston and Marián Janiga. The title is *Feral Pigeons,* that being Johnston and Janiga's careful label for the avian pests that waddle through parks and piazzas on five continents, that nest in attics and rain gutters and clock towers, that shit without rancor or partisanship on equestrian statues of the martial heroes of virtually every nation. Superdoves is an alternate label, offered by the two authors with a half-whimsical disclaimer because it's less precise and more freighted with comment. The word hints at a tangle of evolutionary and ecological factors, tracing backward in time more than five thousand years and hearkening forward toward a dreary postmodern future. At the core of that tangle stands one basic truth: Feral pigeons aren't like other birds.

They aren't even like other pigeons. They fly faster. They eat a more diverse diet. They breed earlier in life, more abundantly throughout it, and repeatedly during the course of a year. They travel long distances, transplanting themselves into new terrain with the robust impertinence of weeds. They have invaded, in particular, the concrete environments that the human species constructs for itself. They succeed in living at high population densities in close proximity to people who despise them. They can hear high-frequency sound; they can see ultraviolet light. They possess an extraordinary degree of genetic variability and an amazingly reliable sense of orientation. Laboratory experiments suggest that they are capable of symbolic reasoning. If those traits don't qualify them as super, what could?

They aren't wild animals and they aren't domestics. They hover around humanity like a guilty memory, flighty but ineradicable. They are genetically designed for survival in the severe urban landscapes of the late twentieth century, and it's a fact worth noting that we ourselves helped design them. Feral pigeons reflect a potent amalgam of two evolutionary forces—artificial selection as practiced by human breeders and natural selection as effected by all other forms of circumstance—that may be unique in the history of animals.

A dour person would say that they're precisely the birds we deserve. Myself, I'm not quite that dour.

THE LONG history of feral pigeons begins with a single species of wild bird, *Columba livia,* commonly known as the rock dove. This was one of five closely related species of ground-frequenting and cliff-nesting pigeons native across large portions of Europe and Asia. It fed mostly on seeds, and it reproduced quickly. Probably it had originated in southern Asia and gradually spread west; fossil remains show that by 300,000 years ago it had reached the eastern Mediterranean. That seems to be where, between five and ten thousand years ago, the rock dove was first domesticated.

Johnston and Janiga suggest that the domestication occurred in two ways, complementary processes with a combined result. First, hunters prowling through rocky terrain must have captured squabs (unfledged nestlings) and taken them into captivity to be reared and fattened as food. That practice would have given way to rearing the captured squabs into breeders and keeping them as a steady, convenient source of meat. The second track to domestication was that some wild rock doves "found the mud and stone walls of human settlements suitable for nestsites and the wheat and barley these people grew suitable for nutrition." So the doves voluntarily took up residence in the settlements, working the market squares and the granaries for what food they could snatch. Such birds were *synanthropes,* in Johnston and Janiga's terminology—that is, born wild but inherently predisposed toward associating themselves with humans.

Populations of feral pigeons presumably became established in early settlements and cities not long after domestication began. Those ferals most likely carried mixed ancestry—some blood from escaped domestics and some from synanthropic colonizers. Seven or eight thousand years later, the phenomenon continues, with one difference—nowadays there are few remaining source populations of wild *C. livia* providing synanthropic colonizers. The wild rock dove survives only on certain northwestern European islands (the Shetlands, the Orkneys, the Hebrides), in patches along the Mediterranean coast, in some mountainous areas of the Near East, and in scattered remnants elsewhere. Even those few populations are presently being compromised by the further geographical spread and genetic intrusion of feral pigeons. That's a whole other side to the feral pigeon story—but a significant one, and we'll come back to it.

With the wild populations so scarce and marginal, new recruits to feral populations come nowadays from just two sources: reproduction by the ferals themselves and a continuing trickle of escapees from captivity. Reproduction by the ferals yields young birds reflecting a strange, mean-streets version of natural selection. And the escapees bring their own genetic baggage, some of which is pretty weird too.

Besides its adaptability to rock-like structures and its willingness to eat grain, *C. livia* carried one other trait that made it a prime candidate for domestication: prolificness. A pair of rock doves would produce at least two or three clutches a year, maybe five or six clutches a year in ideal conditions. No wonder that by about 4000 B.C. pigeon images were appearing in the fertility art of Mesopotamian cultures.

Later the birds came to be valued for more than just food and religious imagery. The practice of breeding them selectively into specialized lineages—for beauty, for whimsy, for show, for sport—probably dates back to the Romans, and it was well established in the Near East by the time of the Crusades. Artificial selection practiced on pigeon stock by fussy and imaginative breeders is what stretched the genetic potential of *C. livia* in various directions. And thanks to the isolation of captive pigeon stocks, one from another, that fabulous diversity was maintained over time within the countless individually managed populations.

"The Arab civilization was the first culture we have record of that kept recognizable breeds of pigeon," according to Stephen Bodio, a writer and naturalist whose own lifelong affection for domestic pigeons and their history is recorded in a wonderful book titled *Aloft.* By 1327, Bodio writes, the city of Modena in Italy had codified guidelines about competitive pigeon-flying. Another place where the mania for pigeons took hold was England. In 1676 an ornithological compendium by Francis Willughby described seventeen domestic breeds of pigeon, notable among which were the tumbler, the carrier, the narrow-tailed shaker, the Barbary, and the Jacobin. The tumbler was a breed that actually turned somersaults in the air as it flew. The Jacobin of Willughby's era, if it resembled the Jacobins of today, was a bizarre thing whose entire head was buried inside a mane of overgrown feathers, so that it could hardly see well enough to feed itself. Francis Willughby's book signals the fact that, already by three centuries ago, the breeding of fancy pigeons had been taken a very long way.

During the late 1850s, Charles Darwin used the selective breeding practiced by pigeon fanciers as an important paradigm for his notion of natural selection, and he included a section on domestic pigeons in the first chapter of *The Origin of Species.* By that time he had joined two of the London pigeon clubs and done a fair bit of pigeon breeding himself. "The diversity of breeds is something astonishing," Darwin wrote. Ten years later, in *The Variation of Animals and Plants under Domestication,* another book exploring the mechanisms of inherited change, he devoted two full chapters to pigeons.

Bodio, in *Aloft,* also remarks on that astonishing diversity. "Only three domestic animals show a comparable range of physical types to the pigeon: the dog, the chicken, and (maybe) the goldfish." In addition to its variability, the pigeon embodies other characteristics that make it appealing to avocational breeders—its early maturity, its fast-hatching eggs, its year-round reproductive activity—because they help shrink the time in which a breeder can see results. Besides these, Bodio cites one further attraction: "Pigeons can *do* various things, unlike a goldfish or (unless you are a cockfighter) a chicken." One of the things they can do is fly home fast.

Racing homers, originating around 1800, were quite different from the Jacobins and fantails and frillbacks and pouters and all other fancy breeds. Racing homers were bred for performance, not for show. Their inherited compulsion and talent was to fly fast and return home reliably, each to its own familiar loft. One formidable strain of racing homers came out of Belgium, a second strain out of England. In the early nineteenth century, before telegraph and telephone, homers brought stock-exchange information to Antwerp from brokers in Brussels, Paris, and London. In the later wars of the century and on through World War I and even World

War II, they carried tactical messages. And among passionate racing enthusiasts like Stephen Bodio's father, on the outskirts of Boston in the 1950s, they were bred strong, sleek, and fast enough to make 600-mile homing sprints in a day.

Occasionally, though, a homer failed to return home. Maybe it had been killed by a bullet or a hawk; maybe it had lost its way; maybe it had simply gotten distracted or seduced into another path of life and ended up on the streets of a nearby city, like a runaway teenager, scuffing out its sustenance among a flock of tough and enterprising companions. To the pigeon breeder, that sort of errancy was the unavoidable cost of doing business. To the pigeon, it was a challenging reentry into the realities of natural selection.

So PIGEONS have gotten around, and not just by wing power. The appeal that they carry among fanciers has been their ticket to worldwide distribution. They're in Australia, they're in Hawaii, they're in the Arctic, they're in Tierra del Fuego. They have even arrived in Montana.

On a brisk afternoon with snow on the ground, two thousand miles west of the west end of 46th Street, I help Steve Bodio carry water out to the pigeons in his loft. It's a modest plywood shed divided into three rooms, each room furnished with nest alcoves and perches. At full capacity, as it is now, it holds forty-some pigeons. Although his pigeon-fancying tastes are broad, time and space impose limitations, and in this loft Steve is keeping just three breeds: Spanish pouters, English carriers, and racing homers.

The pouters are elegant birds, each of them capable of inflating its crop and hoisting itself into a proud, chesty posture, like a bodybuilder at flex in an overly tight truss. The carriers, despite their name, are also a show breed, with decorative wattles across the top of the bill and around the eyes. "I like breeds," Steve says forthrightly. "I'm not one of these animal rights people who thinks breeding things for characteristics is wrong." From all around us, as he speaks, come the sounds of cooing adults, peeping juveniles, wings flapping as birds resettle themselves. "On the other hand, I'm more interested in the whole gestalt of the breed—how they act, everything—than just in breeding a perfectly round wattle."

My attention gravitates to the racing homers. I comment that they look like rock doves, with their pale gray wings, their dark bands, their highlights of shimmery green. Yes, but only superficially, Steve says. His breeder's eye sees nuances that I don't. Wild rock doves have a steeper forehead and a thinner bill, he explains. Also, these homers are longer legged, more deeply muscled, more streamlined than a rock dove, and half again as big. "They've evolved toward—or, been humanly caused to evolve toward—a fairly efficient so-called wild type," he says. "But with actually a little more efficiency for pure flying than even the rock dove." Besides greater speed, there's another thing. "They're a little more enduring as a long-distance flyer. I mean, what reason would a rock dove have to fly six hundred miles?"

The racing homers, superdoves in their own right, are the breed from which feral pigeons are mainly derived, at least in America. They have probably been a major source of European feral populations too. Those AWOL homers have brought their speed, their endurance, and their directional savvy to the gene pools of feral populations, no small contribution toward making the urban superdoves what they are.

I ask Steve Bodio, breeder of elite pigeons, historian of the fancy, what he thinks when he sees feral pigeons on a city street. "I think they're survivors. I think they're great," he says generously. He appreciates any form of wildlife, he adds, that "treats human edifices as just another part of the environment. They're just living in the human canyon."

THE HUMAN canyon. Not being descended from cliff-dwelling ancestors myself, I can only tolerate it for short visits. After a few days of pigeon watching and other New York business, I catch a cab for the airport.

It's early Saturday morning and, as the taxi heads east along 46th Street, I find midtown Manhattan blessedly deserted. The sidewalks are empty, except for pigeons. On the corner of 46th and Park Avenue, one bird pecks its way sedulously across the concrete, as though expecting at any moment to find a kernel of wheat. At 59th Street, where we turn east again, I notice two dozen pigeons perched along the horizontal bar of a streetlight, like a rank of cadets calmly awaiting the next war. I think of Johnston and Janiga's book, so thorough, so authoritative, so new that even Steve Bodio hasn't yet seen it, and of its comment about superdoves.

The two authors refer passingly to the beleaguered status of *Columba livia* in the wild. They note that genetic dissolution threatens those few remnant populations (which in Johnston and Janiga's British-style terminology are Rock Pigeons, not rock doves), and that it's a situation attributable to habitat destruction and urban encroachment throughout Europe and Asia. They offer a sober warning. "Whether or not we have a conservation program for Rock Pigeons, if there is no reduction of the absolute numbers of humans we think the disappearance of ancestral Rock Pigeons could occur as early as in the last half of the 21st century. But," Johnston and Janiga add, in an afterthought that sounds peculiarly cheery, "we would have superdoves in their places."

I find that cold consolation. Like Steve Bodio, I'm glad that at least some forms of avifauna are willing and able to grace the world's cities, and I'm not aghast that feral pigeons don't happen to carry the same array of genes as *C. livia* in the wild. But the feral pigeon is not—repeat, not—a satisfactory substitute for the wild rock dove.

And this relentless replacement of wild populations by feral ones, rare species by weedy ones, inconvenient beasts by convenient ones, isn't limited to pigeons. It's a lamentably broad trend. Humanity is changing the world's flora and fauna—not just extinguishing many species but also transforming those that remain. We're doing it by the force of our ecological sovereignty and by the evolutionary selection (call it natural, call it artificial) that we exercise.

To say so is to open a large topic—too large for the closing paragraphs of a small essay—but I'll open it anyway, in the hope of leaving you with something to chew on. Consider this troubling but real possibility: that the heavy presence of *Homo sapiens* across all the world's landscapes, our irrepressible self-interest, and our well-meaning management decisions may yield a global menagerie of diminished, tractable creatures. Think of "supergrizzlies" in Yellowstone that are too sensible to eat hikers, "superwolves" in Minnesota that are too prudent to mess with cows, "supertigers" in Nepal that feed dutifully on tethered goats for the edification of ecotourists in blinds, "supergorillas" on the Virunga Volcanoes that carry acquired resistance to whooping cough and prefer a PowerBar to a mouthful of nettle leaves, "superdeer" strolling imperturbably through the suburbs, "superginkgo" trees growing from holes in city sidewalks on a diet of carbon monoxide and dog piss, "supermosquitoes" that drink only from hummingbird feeders.

We're headed toward that, and to me it's a dreary prospect. If we come to such a point, with the surviving species (few as they may be) merely cultivated reflections of human dominance, human sufferance, human fancy, we'll have selected away something precious. When the last beasts and the last plants left alive are all just as super as we are, the world will be a crowded and lonely place.

# *Before the Fall*

In the tree-shaded town where I live, it seemed that autumn had come early this year—earlier even than usual for a north-country place where the cold winds begin blowing in mid-August, the cottonwoods turn color not long after Labor Day, and the first heavy snow often puts a damper on Halloween. This year was different. The leaves flushed from their buds during May and then in June, suddenly, disappeared. They hadn't succumbed to the natural rhythm of season. They hadn't gone yellow, fallen, piled up in the gutters as aromatic autumnal mulch. They had been eaten.

A pestilential abundance of small, hairy larvae had materialized like a plague out of Exodus, stripping the trees of their foliage. The scientific name for these voracious leaf-eaters was *Malacosoma disstria,* though few of us knew that at the time.

"Tent caterpillars," said the local newspaper, vaguely but not inaccurately. "Tent caterpillars," said the city parks people and the man at the county extension service, who were answering calls from dozens of concerned residents every day. The radio said "tent caterpillars" too. And so before long we were all out on the sidewalks, saying "tent caterpillars!" back and forth to one another. In the hubbub, we were too occupied to notice that these particular "tent caterpillars" didn't build tents. We weren't interested in such entomological subtleties. What we wanted to know was how we could kill the damned things before they ate all our lovely urban hardwoods down to stumps.

It was an awesome phenomenon, in its own ugly way. Not every tree was left naked, but many were, especially among the old towering elms and ashes that stand along the sidewalks, arching their canopies out over the neighborhood lanes. It happened fast. The caterpillars did much of their feeding under cover of darkness, and on those coolish June nights we could stand beneath a great tree and hear the inexorable crackle of tiny jaws: a sound like distant brushfire, but alive. In the mornings, we would find the sidewalks heavily sprinkled with their poppy-seed globules of poop. Occasionally a lone caterpillar would rappel down on a filament of silk and dangle there mockingly at eye level. On a day of chilly drizzle, too chilly for caterpillar comfort, we could spot them hunkering sociably, high up on a trunk or in a limb crotch, about two hundred fuzzy gray bodies in each pile-up. Some of us went away for a weekend, leaving the lawn freshly mowed, all seemingly fine, and came home to find that our trees had been defoliated. We climbed up on step ladders and sprayed the caterpillars with soapy dishwater from spritzer bottles. We dosed them with bacterial mists or nasty long-molecule chemicals, as variously prescribed by the local garden stores. We called in SWAT-team strikes by the men from Nitro-Green. All of these measures seemed to be marginally effective at best and, more likely, just poisonous and futile. The caterpillars continued to chomp. When they began migrating from ravaged trees to healthy ones in search of more food, we wrapped girdles of duct tape around the tree trunks and smeared on barriers of impassable goo. The caterpillars kept coming. We stepped on them as they forded the sidewalks. We mooshed them wholesale in the streets. But there were simply too many, and the infestation proceeded along its natural course. They ate, they grew, they molted repeatedly. They marched across town, treating our trees like celery.

Eventually they finished eating. They had bulked themselves up to the limits of their caterpillaroid juvenility, and now they were ready for puberty. They spun themselves up inside leaf-wrapped cocoons for a short metamorphic respite, to emerge in a few weeks as little brown moths. The crackling stopped and the treetops, what was left of them, fell silent. The caterpillars, qua caterpillars, were gone. But this vast population of pestiferous lepidoptera was still lurking over our heads, literally and otherwise, like a large gloomy hunch about the future.

BIOLOGISTS have a label for such an event. They call it an outbreak. It's characteristic of certain types of animal but not others. Lemmings undergo outbreaks; river otters don't. Some species of grasshopper do, some species of mouse, some species of starfish, whereas other species among the same taxonomic groups don't. Consider reptiles: Under extraordinary conditions, certain snake species (such as the brown tree snake, *Boiga irregularis,* native to New Guinea but inadvertently transplanted to the island of Guam) may outbreak into staggering plenitude; desert tortoises in the American Southwest don't. An outbreak of woodpeckers is unlikely. An outbreak of wolverines, unlikely. Among those insect species inhabiting forests, about 99 percent maintain stable populations at low density and only one percent ever experience outbreaks. What makes a species of insect—or a species of mammal or reptile or microbe—capable of the outbreak phenomenon? That's a complicated question the experts are still trying to answer.

An entomologist named Alan A. Berryman addressed it a few years ago in a paper titled "The Theory and Classification of Outbreaks." He began with basics: "From the ecological point of view an outbreak can be defined as an explosive increase in the abundance of a particular species that occurs over a relatively short period of time." Then, in the bland tone of a careful scientist, he noted: "From this perspective, the most serious outbreak on the planet earth is that of the species *Homo sapiens.*" Berryman was alluding, of course, to the fact that we've increased our population by a factor of five hundred since the invention of agriculture, by a factor of five since the Industrial Revolution, by more than double within only the last century—and that there seem to be no natural limits in sight. Relative to other large-bodied mammals, we're a grossly abundant species in the throes of an exceptional, and seemingly unsustainable, episode of proliferation and consumption. From Berryman's viewpoint, we're the primate equivalent of an eruption of weevils.

It's a provocative idea, vast in its implications, majestically dour, and certainly not unique to Alan A. Berryman. In fact it's nothing less than a theory of human history, human demography, the terminal destiny of civilization, conceived in cold-bloodedly ecological terms. But, having invoked it, Berryman let it drop. He was just a guy writing about insects for an audience of other entomologists. "From the more narrow perspective of *Homo,* however, an outbreak is an increase in the population of an organism that has a deleterious influence on human survival and well-being," he hastened to add. That brings us back, from a brink of ponderous speculation, to the tent caterpillars in my little town.

Bare trees in midsummer didn't threaten our survival but they did vitiate our sense of well-being.

Two species of *Malacosoma* are native hereabouts: *M. californicum,* the western tent caterpillar, and *M. disstria,* known as the forest tent caterpillar despite the fact that it doesn't build tents. The forest tent caterpillar is the most widely distributed species of *Malacosoma* in North America, and by some measures it causes the greatest damage, at least partly because it feeds on a broader selection of tree species than the others. Among the intriguing facts about *M. disstria* is that it's a social species, capable of reciprocal communication and collective behavior that enhance the welfare of the group. Ants, termites, and some species of bees and wasps have traditionally been considered the only social insects. But it turns out that the forest tent caterpillar, though not so elaborately social as a colony of ants or termites, does engage in cooperative activity.

The eggs are laid in single batches, roughly two hundred eggs in a batch, so that all the offspring from any one female begin life together, as an aggregate litter of siblings. The egg mass remains dormant through winter, glued onto a branch with a frothy secretion from the mother. In springtime, about when new leaves are emerging, the eggs hatch and the tiny caterpillars commence eating. As the food resource becomes scarce in their immediate vicinity, they start to move—and they move as a herd. Crawling along a branch, each caterpillar releases a fine cable of silk and a smear of some pheromonic chemical. One caterpillar locates a fresh feeding site, others arrive and join the feast, they all gorge themselves to the point of satiation, and then they withdraw a short distance to rest. Unlike the western tent caterpillar, the forest tent caterpillar does its resting without benefit of a tent. It congregates, instead, in an open-air bivouac. Even without the protection of a tent, this bivouac behavior carries certain advantages over a lone-wolf style of caterpillar self-interest. Huddling together, the forest tent caterpillars help one another stay warm. When a predator threatens, they rear up on their larval haunches and wave the front ends of their bodies back and forth in the air, showing off the long fuzzy hairs that make them unpalatable tidbits and warning their siblings to do the same. Such a tangled mob of waggling, prickly larvae is even more unattractive to hungry birds, evidently, than a single tent caterpillar would be on its lonesome.

The communication that facilitates this collective bonding is done, as in ants, with chemical trails. Two scientists named T.D. Fitzgerald and James T. Costa, working in a lab full of *M. disstria* and cardboard mazes, found that the pheromone smear laid down by one well-fed caterpillar "appeared to stimulate and orient trail following by siblings still at the feeding site, eventually leading the colony to aggregate at the new bivouac." The caterpillars were enticing one another to seek safety and comfort in togetherness.

This may be one of the evolutionary secrets that allows the forest tent caterpillar to flourish, occasionally, in immense numbers. It seems that crowding is good for them, not bad. And in their dim caterpillical way, they know it.

THE WESTERN tent caterpillar, that other species of *Malacosoma* whose distributional range includes my little town, has its own set of adaptive tactics. Each sibling colony erects a tent, instead of relying on the more chancy (but also more economical) method of open-air bivouac. It centers its feeding excursions around the ever-enlarged tent, rather than foraging nomadically over a broad area as the forest tent caterpillar does. Whether the *M. californicum* set of tactics is superior to the *M. disstria* set of tactics is an issue that evolution hasn't decided—except on a provisional, year-by-year basis. In a normal year, both species are present as small, sparse populations of which no one except the most vigilant entomologist takes notice. In an exceptional year, one species or the other explodes to conspicuousness as a huge population—an outbreak. Then, after a couple years of booming, the outbreak goes bust and that species collapses back to obscurity. Further years pass, with no major eruptions of either *disstria* or *californicum,* until there comes a different mix of conditions and suddenly—VWOOM!—the other species commits an outbreak. Why does one set of adaptive tactics work egregiously well in a given year, while the slightly different tactics of a slightly different species produce no conspicuous population at all? I called the county extension service and asked a fellow that question. He had no answer.

So I pestered him with another: Why has *this* summer, as opposed to last summer or the summer before, been such a hospitable season for *Malacosoma disstria*? What determines the timing of these outbreaks? "I don't think anyone can say why you have the boom and bust," the extension man told me. "It just happens."

SOME RESEARCH entomologists have gone into the subject a bit deeper. Judith H. Myers, for instance, at the University of British Columbia, has derived some fascinating suppositions from her twenty years of work on tent caterpillars and other forest lepidoptera. Among all the insect species that experience outbreaks, Myers reports in a recent issue of *American Scientist,* "there are a few that have even more improbable population dynamics: their outbreaks seem to occur on a regular schedule." She calls these the "cycling" species, as distinct from the species (such as grasshoppers) that outbreak erratically but don't show a cyclical pattern. The western tent caterpillar, in British Columbia, has undergone a series of major outbreaks at eight- to eleven-year intervals, as traced back to 1936. The tussock moth, which infests Douglas fir in western North America, and the spruce budworm, on the eastern edge of the continent, are also cyclical outbreakers. So is *M. disstria,* the forest tent caterpillar—with a history of neatly spaced outbreaks in Ontario dating as far back as 1867. And the eight- to eleven-year interval is common to many other cycling species.

Why do the outbreaks begin and end with such regularity? Population levels are affected by a number of factors, Myers writes, but a cyclical pattern seems to imply the predominating control of some *single* factor. What's the factor? That's been surprisingly hard to discover.

One obvious hypothesis is weather. A wet spring with coolish temperatures (as in my town this year) might provide *M. disstria* with ideal conditions and trigger a population outbreak. The outbreak might continue for two years, or three, until some other quirk of weather interrupts it. But the weather hypothesis isn't supported by evidence. Outbreaks of forest tent caterpillar, for example, seem to show little correlation with particular sorts of weather, either at the onset of each outbreak or at the crash.

Another notion—call it the Malthusian hypothesis—is that the outbreaking population eats up all its food resources and then suffers a massive die-off from starvation. This doesn't fit the facts either, according to Myers. The rise and the fall of cyclical outbreak populations don't correspond closely to changes in food supply. "Cycling insects sometimes decline well before forests are defoliated." The crash comes, that is, while food is still available. And when the trees have recovered completely, the insects remain scarce for a number of years before beginning a new outbreak.

Myers dismisses a few other hypotheses before offering one of her own: nuclear polyhedrosis viruses—or, as they're familiarly known in the scientific literature, NPV. The NPV comprise a large group of viruses that afflict many different species of butterfly and moth. Within a population of victims, they pass from one caterpillar to another in the form of crystalline polyhedra, which are swallowed along with leaf matter and then dissolve in the caterpillar's gut to release replicative viral bodies. The viruses multiply first within stomach cells, spread to other body cells, and eventually kill off the caterpillar, which disintegrates fast, leaving behind more polyhedra to be eaten by other caterpillars. When the infection of NPV reaches a critical point throughout an outbreak population of caterpillars—according to Myers's scenario—the population crashes, irrespective of weather or food supply or any other factor. The similar eight- to eleven-year patterns among many species of cycling insects, Myers guesses, might reflect their similar susceptibility to this one group of viruses.

And the nomadic behavior of *M. disstria* in particular—moving often, foraging widely, leaving behind uneaten leaves, and not confining itself within unhygienic tents—might represent an advanced adaptation to help forest tent caterpillars escape from their own plagues of NPV.

THE LILACS were spectacular this summer in my little town. None of us could recall a year when the hedges had been more thick with those lavender blossoms, when the evening breezes of June had been more rich with that hopeful perfume. It was a great year for raspberries too. We gobbled them by the handful and we froze them by the pint. The bushes began yielding around the middle of July, and a month later they were still heavily laden. Uneaten berries were left to rot on the plants; ripe berries fell into the lawn furniture, causing red stains but making us feel wealthy and blessed. The wet cool spring had given way to a season of thunderstorms, punctuated sparsely by days of sunshine, so that the grass grew long and thick and remained verdant well into the back half of August, even without a fastidious wrangling of sprinklers. That seemed miraculous. And we all heard the scuttlebutt from the countryside beyond the town limits: a dismal year for hay (since there were never enough sequential sunny days to get it cut, dried, and baled) but a helluva fine year for wheat. Furthermore, the trees that had been stripped here in town—instead of dying as we had feared—put out a new growth of tender young leaves. It was a second spring, out of season but nice.

The tent-caterpillar cocoons released adult moths and then drifted down, empty, like cotton balls. I collected half a dozen in the course of one morning's walk and saved them in a jar. I also captured a single brown moth. The other moths had mated and, no doubt, laid their eggs again in our treetops. But that business was done inconspicuously. I continued thinking about *Malacosoma disstria,* though it had reentered the invisible phase of its cycle and disappeared from the trees, from the sidewalks, from the newspaper. I wondered: Would the caterpillars return in such abundance next year? Maybe their success had been fleeting and rare, made possible only by the same quirks of weather that were a boon to the lilacs and the raspberries and the wheat. Maybe they had succeeded too well, multiplied too grossly, and next year they'd come up against starvation. Or maybe they had already fallen ill.

Some days I sat in a lawn chair beneath the apple tree, contemplating tent-caterpillar ecology and reading about the demography of *Homo sapiens.* Specifically, I turned to a handful of scientists who have considered the cataclysmic growth of human population, and its possible consequences, from an ecological and evolutionary perspective. I read a gloomy but plausible book titled *Overshoot,* by William Catton. I read *The Fates of Nations,* by a brilliant ecologist named Paul Colinvaux. I rummaged through *And Replenish the Earth,* by another respected ecologist, Michael Rosenzweig. I dipped back into the works of Paul Ehrlich, Julian Huxley, Garrett Hardin. I checked the indices of their books for the words *outbreak* and *crash.* I found a welter of carefully argued ideas, but not the obvious one I was searching for.

Therefore this is a question, not an assertion: Is *Homo sapiens* an outbreak population, just reaching the peak of its curve, as the entomologist Alan Berryman suggests? And if we *are* presently experiencing an outbreak, what does science warn us to expect?

The raspberry bushes, twenty feet from my chair, drooped and went yellow with depletion. It was a glorious summer. Now comes the fall.

[III]

# *The* MOUNTAINS

# *Pinhead Secrets*

A tall fellow in a Batman costume comes soaring out of the winter sky, his cape flapping wildly, his skis showing me their bottoms. He has hit the jump nicely and gotten bat-worthy air. His eyes, magnified by goggles and framed by his silky black cowl, appear wobbly with mad abandon. His skis yawn away from his heels alarmingly. Has something gone wrong with his equipment? No, it's just that he's wearing telemark gear—that is, flexible boots and free-heel bindings, designed for that style of skiing wherein the turns are performed with an uncanny kneeling motion, and stability at high speeds seems a form of magic. Magic or not, for a moment it looks like this Batman may tumble forward into a great snow-eating kablooey on the lower slope. But again, no. Arms wide, cape deployed, he drops one knee on landing and screeches niftily to a halt.

An envious thought fills my brain: Now *that's* skiing. Then I recognize the caped commando behind his goggles and cowl. It's Noah Poritz, a freelance entomologist who runs an integrated-pest-management business down in town. The cowl, I notice now, includes a pair of pointy black ears. If a costume is worth doing, I suppose, it's worth doing well. Hi, Noah, I say. How's the racecourse?

The gates are fine, he says, but this second jump probably needs to be higher.

Easy for him to say, with his preternatural powers of balance and flight. For me it's another matter. My telemark technique is shaky, my experience is minimal, I'm not wearing a cape, and the second jump certainly *doesn't* need to be higher.

Despite these misgivings, I find myself already infected by the same manic energy that has brought Noah and a hundred other costumed skiers out on this blustery Saturday in late March to the 13th Pinhead Classic, a fervently unserious telemark race held at Montana's Bridger Bowl ski area. "Free your heels, free your head," is the telemarker's apothegm, and the Pinhead Classic traditionally reflects that, marking the end of the season in a spirit halfway between Halloween and the Harlem Globetrotters. There are some masterful skiers in the field for today's event, but the main issue won't be *How good was your time?* The main issue will be *How free were your heels?* Prizes are promised in a wide range of categories—best crash, biggest air, best monster, best nordic attitude, oldest skier, politically incorrect, and leather-and-lace, among others—and the dreary numerical record of elapsed milliseconds between start and finish will be given only perfunctory attention. As I ski down to register and add a race bib to my getup, I'm comforted by a sense that, however ridiculous my own efforts at slalom-and-jump telemarking may prove to be, amid this festival of the ridiculous no one is likely to notice.

MAYBE YOU know the feeling: You sit on a ski lift wearing iron-maiden boots clamped into lock-down bindings while, far below, a figure in free-heel gear and backcountry clothing genuflects through a series of graceful, controlled reverses on a steep slope. And you wonder: How does he do that? For a dozen years, on the lifts above Bridger Bowl and elsewhere, I wondered the same.

The source of this wonderment is a mysterious little maneuver called the telemark turn. Ski books with a sense of history tell us that it was developed by a jumper named Sondre Norheim, from the province of Telemark in southern Norway, and that Norheim unveiled it at a competition in Oslo in 1868. This suggests that the telemark position was a ski-jump tactic initially, adding fore-and-aft stability to landings, and that only secondarily did it reveal its utility for turning. During much of the following century the telemark principle seemed almost lost to the civilized world—like the Dead Sea Scrolls or the Gnostic Gospels, only more fun—while alpine (that is, downhill, in the familiar sense) skiing took shape as a popular sport. Within that alpine tradition, boots and bindings became ever more rigidly Cartesian, holding the entire foot immobile relative to the ski. Vise-like contrivances replaced straps, buckles replaced boot laces, plastic replaced leather, while rope tows and then chairlifts compensated for the fact that a person couldn't *move* in these things unless pointed downslope. Meanwhile the telemark turn wasn't really lost. It had only been hiding in the backcountry, where it was put to good use by nordic (that is, cross-country) and mountaineering skiers who needed the freedom of light boots and unattached heels for covering up-and-down terrain.

As practiced on long wooden touring skis with lignestone edges by five generations of cross-country cruisers, though, the telemark turn was an iffy means of changing direction. Here again I speak from experience. Even my patrilineal roots, which happen to be traceable to the province of Telemark, didn't guarantee me reliable results. The telemark turn as I knew it for decades had a swan-like majesty when it worked and a grandiose comic futility when it didn't, which was often. Generally it was executed with overextended legs, waving arms, and a facial expression approximating Munch's "The Scream," in which form it afforded all the stability of a bicycle ridden too slowly along a high girder. It could be useful on gently sloped meadows of knee-deep powder, and with twenty or thirty yards' leeway it might barely deliver you from the path of an oncoming tree. But the notion of adapting it to serious downhill skiing— say, telemarking through gates on a hard-packed intermediate run at your local ski area—was virtually unthinkable. Two factors combined to change that, yielding the modern telemark revolution: better technique and a new class of equipment.

Both these waves of innovation began to show in the 1970s and have crested steeper just in recent years. The new equipment included sturdier boots, sturdier toe-pin bindings, adjustable heel cables giving still more control, and specialized skis with the right sort of edges, the right sort of camber, the right sort of width and sidecut for applying the particular dynamics of the telemark turn on a variety of snow surfaces. The better technique became available as a small number of pioneering skiers discovered for themselves what could really be done in the telemark mode, and as a handful of writers, teachers, and video makers explained the mechanics of those discoveries for the rest of us. Two fellows whose names would become resonant in telemark history, Dickie Hall in Vermont (see "The Big Turn") and Rick Borkovec in Colorado, were especially influential among the pioneers. Paul Parker, a longtime ski teacher and equipment designer, has presented a touch of historical background and an abundance of lucidly explained technique in his excellent book *Free-Heel Skiing*. Bela and Mimi Vadasz have run one of the most widely respected schools for free-heel technique, Alpine Skills International, from an old lodge near Donner Pass in the Sierra. These folks and others have brought the telemark turn out of the backcountry and onto the expert slopes at places like Crested Butte (Colorado), Mad River Glen (Vermont), Steamboat Springs (Colorado), Sun Valley (Idaho), and little Bridger Bowl, at the last of which I myself watched it with growing fascination, wondering whether I too might be capable of achieving snappy turns, high speeds, centrifugal thrills, and severe bodily injury on telemark skis.

My recent research suggests that the answer is yes.

In early January, I began again as a novice: learning to ski in a new way on a new kind of gear. Telemarkian genes would count for little or nothing, I knew, but maybe I wasn't too old to be educable. I bought telemark skis, telemark boots, telemark bindings, and a copy of Parker's *Free-Heel Skiing*, which became my first teacher.

A couple hours with the book gave me more insight into telemark technique than I had gotten by trial-and-error during the previous twenty years. The barest essentials came as epiphanies to me, and I found myself highlighting them in pencil. "Sink *between* your feet, with both skis equally weighted," Parker advised, correcting my notion that the front ski received most of the weight. "Be sure to avoid a too-wide stance," he said, as though he had been watching all those times I tried to do the splits. Exert simultaneous and complementary pressure with your big toe (on the front foot) and your little toe (on the rear foot) to set the edges of your skis for a turn, he explained. And don't—repeat, don't—let that rear foot dangle daintily. "Keep your entire forefoot on the ski," he insisted, so as to maintain your weight load and prevent the rear ski from flapping around like an empty U-Haul trailer in a high crosswind.

In addition to these particulars, Parker offered a general tip. "Each winter I set a goal for my skiing. That goal might be skiing a certain tour, a special race, or descending a peak or couloir I've been eyeing." Every reader should do likewise, he suggested. It seemed reasonable enough. My own goal, I decided, would be to acquire enough skill and confidence within ten short weeks to make an utter fool of myself at the Pinhead Classic.

My first moments of experiment with the new gear, on an easy slope at Bridger Bowl, brought an astonishing revelation: Yes yes, I could turn these skis. The hill was hard-packed but the steely edges and the sidecut provided vastly more grip than I had expected, and the boots were just stiff enough to communicate my intentions. When I genuflected to the snow god, he blessed me with an almost miraculous change in direction. After a few minutes of reveling in this discovery—right turn, left turn, back and forth—I prostrated myself reverently to the hibernal deity. In secular terms: I slammed my face into the snow.

Even the fall seemed encouraging. Someone had told me that, as a telemark beginner, if I fell backward or into the hill, I was doing it all wrong, whereas if I fell downhill on my belly, packing snow into my eye sockets and my ears, ramming snow up my sleeves, leaving crash-dummy impressions of my profile in the pavement-hard surface of the piste, I was doing only part of it wrong. By that standard, I made respectable progress on the first day. My turns were slow, unsteady, imperfect, but roughly telemarkish; my falls were exemplary.

Over the following weeks I got back onto the hill a dozen more times to practice what I'd gleaned from Parker's book:

weight on both skis, stance compact, big-toe-and-little-toe pressure, hands forward. But the book, for all its merits, was just a book. So in mid-March, feeling hungry for face-to-face instruction, I packed up my tele skis and headed to Alpine Skills International. It was time to submit myself to the pitying glare of experts.

"TRUTH IS, the telemark probably isn't the most efficient way to turn," said Mike Carville, a lean 31-year-old rock climber who spends his winters teaching at ASI. "It's probably not as quick, edge to edge, as an alpine turn. The stability isn't quite as good." And the strenuous up-and-down motion of telemarking burns up your thigh muscles more quickly, Carville said. Then he paused, while a smile like winter sunshine lit his face. "But, God, it's fun to do."

Alpine Skills International is an informal but fiercely competent operation based in a boxy building on old Highway 40 where it crests the Sierra, not far above the suburbs of Lake Tahoe. Bela Vadasz and his wife Mimi founded the enterprise in the late 1970s and moved into the present lodge in 1983. Just a short walk away is Donner Ski Ranch, the small lift-served area where much of ASI's instruction takes place. Mike Carville had drawn the chore, this weekend, of teaching a seminar labeled "Telemark and Nordic Downhill" for eight eager and ungraceful beginners, including myself.

The seminar title, with its seemingly oxymoronic phrase "Nordic Downhill" combining what are ordinarily considered two antithetical ski traditions, reflected a fundamental aspect of ASI philosophy: that the telemark turn is simply one tactic among many that a complete skier can use for exploring the mountains. That philosophy lay behind the wide range of offerings in the ASI catalogue, from introductory cross-country to advanced randonnée touring to ski mountaineering to steep-chute alpine. Bela alluded to it when I asked him the same dunderheaded question I'd already asked Mike: Is the telemark turn an end in itself, or is it a tool?

"I think of it as a tool. As part of the total bag of tricks," he said. "I'm not hung up on the telemark turn. I'm hung up on skiing." When the snow is good, he explained, a parallel turn, even on free-heel gear, may be preferable. But on the sticky snow of a warm afternoon, or when you need additional fore-and-aft stability because you're carrying a pack, you can switch to the telemark. Your free-heel equipment gives you the mobility to escape beyond the lift-served areas, and your mastery of both telemark and alpine technique gives you the versatility to ski whatever comes. Having said that much, though, Bela allowed that the telemark turn has an intrinsic satisfaction that can't be measured in terms of practicality.

I had heard this theme before and I would hear it again: that the ultimate logic of telemarking—insofar as it has any or needs any—is existential. Practicality is the just the facade for its heart-gladdening impracticality. Skiing itself is an impractical activity, after all, although some modes (old-fashioned jumping, slambam mogul runs, maniacal leaps off high cornices) are more extravagantly impractical than others. With telemarking, the special existential pleasure isn't measured in dimensions of danger or speed or height. It pertains more to rhythm and grace, to fandangos of deftness, to the challenge of performing a complicated feat with simple equipment. Making a telemark turn, like casting a tight loop with a fly rod (which is not always the most efficient way to catch trout) or carving a controlled path through a class-five rapid in a kayak (which is not always the most efficient way to avoid drowning), has a rationale that transcends practicality: It feels beautiful to do.

Two days with Mike Carville gave me a much better attunement to that dimension. Part of the seminar's merit lay in what Mike himself showed us about telemark technique; part emerged from the fact that we were a group of struggling peers, watching one another make mistakes that were congruent to but more easily recognized than the mistakes we were making ourselves. By the end of the first day I had fallen into friendly alliance with three other students who were progressing at roughly my speed: an astronomer named James, a physician named Kathryn, an environmental engineer named Simon. All three happened to be young, professional, and from the Bay Area, but otherwise we had plenty in common. For instance, we each tended to underweight the rear ski. We each displayed an amateurish lack of angulation—that is, angling the lower body and the ski edges into the slope, while keeping the torso erect and dipping the downhill shoulder. We each suffered from drifting hands that led often to waving arms. And James in particular executed some splendid face-smashing falls, one of which was captured on video for us all to review and admire back at ASI.

Skiing hard for two days together, with the benefit of expert Carvillean commentary, James and Kathryn and Simon and I could actually see one another improving in brisk strides. We began to link turns in steady rhythms. We grasped the importance—much harped on by Mike—of driving the uphill hand out across the downhill ski, and thereby keeping the torso squared to the fall line. Soon we were swooping down the steepest slopes (which weren't all that steep) on Donner's sun-blasted back side. When we fell, we fell forward, testifying that our wrongness, however emphatic, was partial. Then the seminar was over, the other members of my telemark assault team went back to their jobs in the Bay Area, and Mike Carville turned to other tasks. As for me? I was ready to go home to the Pinhead Classic.

THE CONVENTIONAL notion of telemarkers—that they represent only a minuscule, stubbornly unfashionable, and historically retrograde group of granola-heads within the larger community of skiers—is enticing but wrong. The reality is that this branch of the sport is growing at a modest but significant rate. According to industry figures, roughly twenty thousand pairs of metal-edged nordic skis suitable for telemarking and mountaineering are sold annually. Telemark equipment now accounts for about ten percent of the cross-country-ski business, and while other sectors of the cross-country market were slumping in recent years, telemark held steady. Dickie Hall's loose-jointed enterprise, the North American Telemark Organization, teaches telemark technique to about 5,000 skiers each year, and Hall himself has taught about 30,000 telemarkers and 1,000 instructors. The national totals are considerably larger, but telemarking doesn't seem so conspicuous on the downhill slopes as, say, snowboarding now is, because many telemarkers still do their skiing in the backcountry. Where do the new telemarkers come from? About half are alpine skiers hankering for a fresh challenge, Hall says; the other half are cross-country skiers who want the enhanced control and the more intense thrills of telemark technique. Very few people hit upon telemark as their entryway into skiing. But the switch from alpine or cross-country to telemark tends to be gratifying and permanent, as Hall's informal polling indicates. He encounters huge numbers of cross-over skiers every year, he says, but seldom meets anyone who "used to be" a telemark skier.

Racing on telemark skis attracts fewer people, but still a much broader and less elite segment than the phrase *ski racing* might suggest. The history of telemark racing in the United States is short but vivid, traceable to Colorado in the mid-1970s. There were early races at Breckenridge, Crested Butte, Steamboat, elsewhere in the state, and then before long also in other parts of the Rockies and at certain telemark strongholds in New England. The U.S. Ski Association recognized telemark as a legitimized genre of competition, and results from the regional series counted toward qualification for a national championship. Russ Tuckerman, Keith Talley, and Whitney Thurlow were three young Montana telemarkers who, during the 1980s, raced seriously and took part in the nationals.

Some years they would rent a Winnebago, drive down to Crested Butte or wherever the nationals were held, park the motor home in a snowdrift, and live out of it while they competed. Tuckerman eventually raced internationally with the U.S. team, while working as a sales rep and a backroom designer for the Fischer ski company. Talley returned to Bridger Bowl as a ski instructor with a special felicity for telemark. Thurlow, vastly talented, became a four-time national champion. Meanwhile the telemark racing scene came to a crescendo and then, according to Tuckerman, changed. Some telemarkers found themselves afflicted with competitive burnout; others, with organizational burnout. The USSA withdrew its sanctioning. The burned-out competitors reverted to backcountry skiing and melted away into the woods or, if they did continue racing, found themselves more comfortable with what Tuckerman calls "the festival atmosphere." That festival atmosphere entails a more frivolous, more inclusive, more hedonistic, and less rigorously competitive approach to race events. The Pinhead Classic is a good example of the festival atmosphere run benignly amok.

Tuckerman, Talley, Noah Poritz, and a few other skiers have played principal roles in staging the Pinhead annually at Bridger Bowl and in setting its tone. Now in its lucky thirteenth year, it's the sort of event for which a former international-class telemarker might turn up in a gorilla suit and cheerfully take his place in the paired starting gates beside some clumsy doofus like me. Informal as it may be, the race isn't carelessly named. Besides that double-edged *pinhead,* proudly embracing the notion of small hat size among those who ski on toe-pin bindings, the word *classic* itself carries specific meaning in the context of telemark racing. As Russ Tuckerman explains it, the Classic is one of three events (dual slalom and giant slalom being the others) of which a sanctioned telemark competition in the U.S. usually consists. The Classic course, longer and more complicated than the slaloms, with a bit of steeplechase flavor, is designed to test the endurance and the acrobatic abilities of the skiers. In addition to giant-slalom gates, it includes *reipelykkje* gates (banked 360-degree loops), an uphill stretch paying tribute to the nordic roots, and one or two medium-sized jumps. Apart from there being no uphill stretch, that's just the sort of ugly, sardonic juggernaut that I find myself facing here at Bridger, now that my crash course of training is done, my entry fee is paid, my race bib is tied in place over my costume, and I'm standing near the starting gates, surrounded by dozens of amiable people disguised as nutcases.

Russ Tuckerman is dressed as Sondre Norheim, it seems, in a set of plaid flannels and a pair of jumping skis so long and wide that they might be landing gear for a bush plane. Another racer, in a baggy brown business suit and a Ronald Reagan mask, shares jellybeans with the rest of us as we wait. Someone is costumed as a dinosaur, someone else as a pig, someone else in a poof-sleeve Elizabethan tunic and tights, someone else as Munch's screamer, someone else as a duck blind with a mallard decoy for a hat, someone else as a ballerina in Day-Glo green. There's a fellow in a zoot suit, another fellow in cowboy boots and a lace nightie (contending for the leather-and-lace prize, evidently), a third fellow in blackface and drag as Aunt Jemima (politically incorrect), and still another cross-dressing man in a svelte gown. Two women have paired themselves as the figure-skating stars Tonya Harding and Nancy What's-her-name (Tonya resplendent in a trashy black tutu, with a plastic baseball bat and a broken boot lace), and two guys have teamed as Beavis and Butt-head, except that today Butt-head is Pin-head, wearing a safety pin the size of a coat hanger stuck through his temples. Need I reiterate that all of these people are on skis? There are racing cars made of cardboard and worn on suspenders. There are costumes from the thrift shop, from the hardware store, and several apparently from Frederick's of Hollywood. Noah Poritz is here as Batman, and his wife, Leona, as Batwoman. The weather has stayed cruel, the wind is sharp, the air is full of driven March slushflakes, and we're all festively freezing our buns. Keith Talley is costumed as a sensible skier in a warm jacket with a clipboard, an excusable breach of style since he's running the starts. And unfortunately now, speaking of starts, it's my turn.

In this first run I'm matched against Shona the Flower Lady, a young woman attired in a long fluffy dress and an Easter bonnet, her poles festooned with flowers. My toes have gone numb during the wait, but they seem to thaw instantly when Shona and I break the start barriers. I round the first couple gates in a recognizable approximation of telemarking, and I loop through the reipelykkje without much trouble. Then I snowplow cravenly into the first jump so as to minimize my air, although that doesn't prevent me from falling. It's okay to fall at the Pinhead Classic—especially okay on this run, since in my peripheral vision I've seen Shona fall too. I recover quickly, make the rest of the gates, creep over the second jump, and then tuck toward the finish, but by that time Shona is well ahead. Either she skied a great lower half or she elected to bypass a few gates and go for raw speed.

As we gather back at the top for our second runs, some generous soul passes a mason jar full of moonshine down the line. It tastes horribly good and helps rekindle our internal fires. Now the journalist in me asserts itself, and I conduct a brief attitudinal survey, working my way up the line to ask racer after racer: *Why?* Why does a person dress in a silly outfit and clamp on unsteady skis to run gates on such a sublimely miserable day?

"I don't know. 'Cause I'll be forty next year."

"It's like being on a different ride."

"I'm a little bit insane."

"Pandemonium."

Behind me, as though on cue, a racer breaks from the start to raucous hoots of encouragement. Having stripped off his trench coat with the countdown, he's already halfway to the second gate in his Jockey shorts.

"Because it's fun. It's a tradition."

"None of your fucking business."

One fellow gestures mutely at the sky, at the ridge, at the whitened slopes, at the Montana panorama: *That's* why, *that's* why.

"Because it's there," the pig tells me. Then he stops himself. Remembering his free-heel bindings and the whole ethic of revolt and relinquishment that attends them, he makes a correction. "Because it's *not* there, actually." He adds: "It's a way to be. And we all want to be in that nordic mood. Free. Free-heeled. Free-wheeling."

Shona the Flower Lady agrees. She was an alpine skier since childhood, she says, but always in pain, tormented by constrictive equipment. No more. "Free your heel," Shona advises, "free your mind."

Late in the afternoon, with the second runs all finished, I find myself perched for a third time at the start. The wind is now pushing a blizzard into the mountain, visibility is poor, and most of the other racers are down in the lodge; but the spirit of the event has captured me, and I can't resist. After some mutual goading by my partner for the day's foolish work, an eminent magazine photographer named Gordon Wiltsie, the two of us have decided to make one further run. This will be the Aging Journalists' Heat, unofficial and untimed. He's handicapped by a pack full of fancy cameras; I'm handicapped only by cold toes and incompetence. How are the jumps? asks Gordon, who has been working so hard all day, shooting film at the starting gate, that he hasn't yet ventured down the course. The jumps? They're fine, I say, having no intention of going anywhere near them this time if I can help it.

We breathe deeply, we crouch, we lean forward over our poles, doing the things we've seen ski racers do on TV. Keith Talley gives us a countdown, and we go. Gordon, being a much better skier, pulls ahead immediately. Later I'll hear how he hit the first jump at full speed, thanks to my assurance that it wasn't problematic; how he flew off into the ether; how he performed a great snow-eating kablooey on the lower slope. But I don't happen to be in position to witness Gordon's crash, even peripherally, due to a mishap of my own.

Relaxed and aggressive out of the start, I catch an edge in the early seconds—all right, I'm just rounding gate one—and plunge headfirst down the hill on my belly. Time suddenly becomes dreamlike and slow, as time tends to do in such cases. Sliding onward, I raise my head to see the paired double poles of gate two looming ahead. And at that instant a thought enters my mind. It's a peculiar thought, witless and illogical, rising from some inchoate perception of unshirkable challenge: With a little rudder work, says the thought, I might nail that gate like a human torpedo.

I gouge at the snow with my hands. I aim myself with the body English of a suicidal skydiver. I assume the unheeding directedness of a bowling ball. Don't ask why I do this because I can't give an answer, but it seems appropriate behavior in the heat of the moment. My aim is pretty good: My head hits one pole and my glasses break in half. Then my shoulders slam tight in the narrow gap between poles, and there I stop. Skis tangled, I roll onto my back. Am I injured? Not badly, it seems. Through a fog of blurred vision and blowing snow, I notice Keith Talley and a handful of other aghast onlookers watching intently to see whether I've cracked my skull. When I raise my clasped hands into a signal of triumph—in the telemarker's cosmos, anyhow, it's construable as a triumph—the crowd goes wild.

# *The Keys to Kingdom Come**

The By-Way Cafe sits at a quiet highway crossroads in central Montana, playing its own small but distinct role in the balance of global thermonuclear terror.

On one wall hangs a Montana landscape, hand-painted on the blade of a crosscut saw. Elk and aspen are featured in a high mountain meadow, though outside the window this part of the state actually runs more to undulant prairie, foothills, antelope, and sage. On the other wall is a rodeo photograph, a moment of bull-riding mayhem captured in black and white, flanked by two of the bad Charlie Russell prints that rural Montanans so dearly love. Nearer the kitchen is a cardboard sign reading, HOMEMADE

*This piece was first published in 1987. Things have changed somewhat along the northern ramparts of the Strategic Air Command, as elsewhere, in the years since. Presumably even the By-Way Cafe has felt the warm breezes of a New World Order, though I haven't been back there to check.

For more information about first publication of—and later chronological adjustments to—this piece and the others, see "Notes and Provenance" at the back.

PIE. Another sign, fixed to the newspaper machine, confesses recklessly: YOUR CONSCIENCE IS MY ONLY PROTECTION. In most respects, for its milieu, the place is quite ordinary. In one respect, no. The By-Way Cafe is a favored caffeine stop for Minuteman missile crews.

"The cinnamon rolls here are legendary," Captain John Cotter tells me.

On this late-winter morning, the room is full of cheerful young men dressed in blue uniforms and strange, colorful scarves. Most are in their middle twenties, the usual age for lieutenants and captains. Most wear eyeglasses. They joke with each other benignly, hollering between tables. Like the By-Way Cafe, they are both ordinary and not. These men are all missile combat crewmen from the 341st Strategic Missile Wing, based at Malmstrom Air Force Base, near Great Falls, and they are all on their way to duty in the buried, shock-hardened capsules from which Minuteman ICBMs are controlled. They will be underground—two men in each capsule, staring at lights and switches, each crew controlling ten missiles—for twenty-four hours. Hence their understandable appetite for last-minute mirth and coffee.

The morning's fresh batch of cinnamon rolls isn't out of the oven yet, but with a soldier's foraging eye, Captain Cotter has spotted some leftovers on a covered plate and requisitioned them for our table. We have no time to wait around for the fresh batch. Cotter and his crew partner, a boyish lieutenant named Ed Burkhart, are due to report soon at the Oscar-1 capsule, where two weary crewmen await relief. Meanwhile, a storm is coming in.

So far there has been only a cold, gusty wind, which blasted against our big diesel van as we drove east this morning into the Judith River Basin. But Oscar-1 is the most remote of all the capsules of the 341st Wing, more than a hundred miles from Great Falls, and we have come only halfway. If a March blizzard appears suddenly and closes these two-lane country roads, the snowbound crew will have to stay on duty until whenever Cotter and his partner arrive. Maybe an extra full day. Possibly two. It's a long time to sit underground, gaping at panels of lights, with a major share of responsibility over the life or death of the world.

Leaving the By-Way, we continue toward Oscar-1, driving a back road that takes us across Wolf Creek and past the big Cargill wheat elevator outside the town of Denton and then down through the stark, water-carved badlands of an area called the Judith Gap. The Judith Gap is a north-south channel between two sets of mountains, notorious for its nasty weather; when a snowy wind blows from the north, the Gap gets strafed. Snow fences line the ridge tops. Ranch houses cower behind carefully planted windbreaks. But the first of Montana's many false springtimes has already melted the hillsides bare, and so far today the wind is just empty bluster. Herefords are grazing placidly. Only the pheasants and the ground squirrels—which would be plentiful along the shoulders of this road on a sunny spring day—have already taken cover. Evidently they know something. We cross the Judith River near a ramshackle ranch house sheltered by cottonwoods, and again the road begins climbing. We talk.

The subject of conversation is, What makes a person become a missile crewman?

"The Surgeon General said I couldn't fly," says Ed Burkhart, and he touches his glasses. "Most of us are frustrated flyboys."

After he joined the Air Force, Burkhart like many of the others faced the fact that imperfect eyesight disqualified him as a pilot. Missile-crew training was another option that offered the chance of an operational (as opposed to desk-bound) career. So he signed up. You don't need twenty-twenty vision to insert and turn a launch key.

That's one explanation, but Burkhart also offers a variant. Behind his Ron Howard wholesomeness lurks a measure of deadpan wit, and he explains that, after getting a business degree at a small Texas college, he worked as a civilian for three years in business management. Then he enlisted. "I wanted to get into a field with a little less stress," he says. "So I went into missiles." Cotter laughs. So do I, nervously.

"Jesus. What sort of business were you managing?" I ask.

"A Radio Shack outlet."

Captain Cotter, on the other hand, is an earnest fellow with a face like an open Bible. Despite his training, despite his job, he seems gentle and likable. He is tall and just a bit gangly, with a vulnerable, goofy sweetness that at times in his life has no doubt been tested by ridicule. He comes from a family in which both parents were teachers and now, while he serves his tour of missile duty, he is finishing a degree in guidance and counseling. After the Air Force, he thinks he might like a career as a high-school counselor—the satisfaction of helping teenagers. With John Cotter as Nuclear Warrior, we enter the realm of oxymoron. He says: "I originally wanted to be a flight navigator. But I busted a test near the very end of my training, and they washed me out." It's a painful admission of a painful setback. Having failed to master the sextant, though, Cotter did just fine in the twelve-week course (for the modernized Minuteman, there is also a sixteen-week version) that teaches young men (and more recently, young women) to launch or not launch nuclear missiles.

Technical expertise, like good eyesight, is not a prerequisite to the job. In just three to four months, a green lieutenant with an ROTC commission and a bachelor's in history or economics can be turned into a missile combat crewman. What the Air Force wants in those capsules are stable, methodical, uncomplicated officers who have been trained to follow checklist procedures and can be relied on, come hell or high winds, to perform exactly as trained.

In essence the job is quite simple: to turn a key when—and only when—so ordered. In practice the launch officers stay busy underground with an elaborate regimen of monitoring and security chores. Still, there has got to be time for boredom. There has got to be time for doubt. Boredom is bad enough, but doubt is anathema, so the Air Force must select these people cannily. The prerequisites for the job are character, of a certain sort, and an unambivalent belief in the moral and practical rightness of American strategic deterrence. Imagination and angst are unwelcome.

I ask Captain Cotter: What sort of young people would *you* choose? What sort of young people would you—as a future guidance counselor—guide toward this particular job? He ponders the question for a moment. "An 'A' student would be nice. But what you really want in a missile crewman, I think," he answers self-knowingly, "is the hard-earned 'B.' "

THE MINUTEMAN missiles of the 341st Strategic Missile Wing hold a special place in world history, because they were the first solid-fuel, silo-stored, and almost instantaneously launch-ready

ICBMs deployed anywhere on the planet. That deployment began in 1962. Ten silos, and the launch-control capsule now known as Alpha-1, just a few miles east of Great Falls, were ready in October of that year, though the ten missiles themselves had not yet been set in place. During the Cuban Missile Crisis, they were rushed into position by assembly crews working overtime.

The Cuban Missile Crisis itself is a useful touchstone, I've found, in discussing nuclear matters with military and weapons-research people. I have a craving to know how other folk remember the experience of that ghastly week, and their various reactions seem to me an important measure of character. If they were scared out of their wits back in October of 1962, I feel more inclined to like and trust them. Captain John Cotter has no memory of the crisis at all, understandably, since he was two years old at the time. Most officers up to the rank of major are also too young, but colonels are just about right. So one evening over steak dinners at the Malmstrom officers' club, I asked a table full of colonels for their Cuban Missile Crisis memories.

The vice commander of the 341st Wing told me he recalled spending the week glued to the television in his college frat house and being agreeably fascinated by the political process in action; this colonel otherwise seemed a sane and charming man. A lieutenant colonel named Jerry Bon, who is John Cotter's squadron commander, told me he wasn't concerned at the time, since he felt confident that Kennedy's resoluteness would bring a peaceful settlement; I envied Colonel Bon his faith. Bon and I were both pubescent high school freshmen in October of 1962, but he evidently suffered a lesser dose of the hormone for nuclear jitters. Another freshman back then was Conrad Strickland—now Lt. Colonel Strickland and Bon's deputy commander. Strickland, in response to my question, acknowledged remembering terror and tears. He remembered walking home from school at the climax of missile week, watching a sunset, and thinking, "This might be the last sunset I ever see." Later there was a dream, a sweat-drenchingly vivid nightmare, in which he watched a mushroom cloud blossom over the American landscape. Only one dream? I asked him. No recurrences? Doesn't everybody have that dream about five times a year? But I was glad to know that it was in him somewhere.

Today Colonel Strickland believes unambivalently in the

rightness of American deterrence, and that the launch keys will never have to be turned.

Each of these colonels is a former missile crewman. They have all sat where John Cotter now sits.

WE REACH Oscar-1 just as the snow is beginning to fly. A sergeant in fatigues comes out to greet us from inside the chainlink fence. Identities are verified, confirming our authorization to enter, and the gate is rolled open. The van passes through, the gate closes behind us, and I am frisked physically and electronically as I stand spread-eagle and coatless in the raw wind. Inside the support building, at ground level over the capsule, Cotter and Burkhart show their ID cards to a team of security police on the far side of a door marked SECURITY CONTROL CENTER, then are admitted through that door. From there they take an elevator down to the Oscar capsule.

In the capsule, they trade code verifications with the outgoing crew. They take custody of two Smith & Wesson .38s. The outgoing crewmen remove their personal padlocks from the door of a red metal strongbox, and Cotter and Burkhart substitute theirs. The contents of the strongbox stay where they are: a set of six-digit Emergency War Order verification codes and two keys.

Cotter and Burkhart begin their shift—in the jargon, they are "pulling an alert." Each sits in a padded aircraft seat before a panel of switches and lights and gauges and knobs. Most of the capsule is filled with computer hardware. It's as comfortable and spacious as the radar room of a submarine. It's furnished with a UHF radio, a keyboard, and a printer. There's also a cot, a telephone, and a television. Yes, they do get the better satellite channels. On the telephone they can talk with a wife or a girlfriend back in Great Falls; they can order a pizza, though for the next twenty-four hours they can't go upstairs to accept delivery. By radio or computer, they can take a message from the president of the United States, who might be ordering them to launch their missiles. If the president ever sends that order, he will need to supply his own six-digit code, accurately complementing the one in their strongbox. In that event, they will insert their keys into the relevant slots on their panel. Of the Smith & Wesson revolvers, John Cotter says,

"I usually hang mine way back on the rack. I don't feel very comfortable around loaded firearms."

One turn of one key does not trigger a launch. The two keys turned simultaneously (their slots are set ten feet apart, so this task can't be performed by one person) don't trigger a launch either. But if Cotter and Burkhart both turn their keys at once, and two other crewmen on alert in another capsule also turn *their* keys, then missiles go. They fly over the pole toward the Soviet Union (or, more recently, perhaps toward China or Iraq; see "Notes and Provenance"). From northern Montana, it's not a long trip. Each missile carries at least one thermonuclear warhead, of which the exact explosive power—say, in multiples of the amount that hit Hiroshima—is classified. Once the four keys have been turned, nothing on Earth can call back those missiles. Once the four keys have been turned, reality and civilization and the logic of American strategic deterrence will nevermore be the same.

Upstairs at Oscar-1, I am eating a cheeseburger and a salad. For this meal the Air Force charges me seventy cents, an unworldly bargain. Soon I'm joined by Lieutenant John Mitchell and Lieutenant Scott Calkins, the missile crew who have just come up from underground. We leave quickly for the drive back to Great Falls, because the weather has turned menacing. The temperature has fallen. The cold west wind is full of snow.

THE MOTIF of the key is everywhere in this business. The key has deep synecdochic resonance—it is both a central reality and a central symbol. Down at Vandenberg Air Force Base in California, where all missile crewmen are trained, the insignia of the squadron that conducts the training is a three-level design consisting of (1) a missile with rocket engines firing, superimposed upon (2) an open book, which is in turn superimposed upon (3) a key. The significance of this design, I've been informed, is that "training is the key to peace." But what I see, of course, is a literal key—a launch key—and a launch. As for the book, I suppose it might stand for Lamentations in the Old Testament, with that opening verse about the fall of Jerusalem: "How doth the city sit solitary, that was full of people! how is she become as a widow!" Or maybe a paperback of *Fail-Safe.*

Much of the training at Vandenberg is conducted in computerized simulators known as Missile Procedures Trainers—MPTs, in the jargon. An MPT is a sealed metal room with two seats and two panels, an exact replica of a launch-control capsule. Besides taking classroom instruction, a missile crewman trainee spends long hours each week in an MPT, along with his or her crew partner (though women are now being trained for this role, the missile crews will not be coed), working through simulated versions of every routine chore and problem, every emergency situation, that they might face in the course of real duty—or at least, every problem and situation that the instructors are able to foresee. A fire in their generator. The trainees cope with it. A disruption of radio communications. The trainees cope. Their blast valves slam closed automatically, sealing the capsule like a can of vacuum-packed coffee, in protective response to some errant Montana crop-duster who has sprinkled their sensors with bug poison. The trainees cope. Their light panels show an Outer Zone Fault, meaning that the security perimeter around one missile silo has been breached; maybe it's a red-tailed hawk, maybe it's a tumbleweed, maybe it's a Libyan saboteur with wire cutters and plastique. The trainees cope. They are graded for the speed and correctness of their reactions. And then somewhere near the middle of this course they receive orders, for the first time, to turn their keys.

The ultimate message arrives. They unlock their padlocks and open the red metal box. They match their six-digit Emergency War Order codes with those supplied by whichever instructor is playing, in this simulation, the role of the president. Confirm validity of EWO codes, roger. We are in a war environment, roger. Missiles have been transferred to *enable* mode, roger. Each crewman takes his or her key, walks back to his or her panel, inserts the key in the slot. On a signal, turns it. Something very important happens next.

What happens is that, having experienced this moment within the MPT, the trainees are immediately invited to sign a statement. The statement is brief and formulaic. It is called the Final Certification of Personal Commitment. It says: "I understand the responsibilities of a missile combat crew member and realize what actions this duty may entail. I certify that I have no reservations over my ability and conviction to perform in such capacity." Yes, if so ordered I *will* turn the key.

Those trainees who sign this statement continue their missile-crew training. Any trainee who cannot or will not sign, and who also cannot qualify as a total (as distinct from selective) conscientious objector, opposed to all forms of violence at all times, can expect to be discharged from the U.S. Air Force. Ironically, a total CO is allowed to remain in the Air Force (for instance, as a medic), but not the trainee who draws a moral line between firing a rifle and firing a Minuteman missile.

That trainee is given an honorable discharge. If the ex-trainee has put himself through college with help from Air Force ROTC, he is obliged to pay back the money.

LIEUTENANT John Mitchell has signed that Final Certification of Personal Commitment.

His crew buddies call him Mitch. Mitch comes from the small town of Limon, Colorado, and he is a good ole boy in a sense of that phrase not meant to be condescending. As we ride back from Oscar-1, crossing the Judith Gap in what is now a full-blown blizzard, our windshield wipers wrapped in ice, visibility coming and going with each shift of wind, Mitch tells me about how he happens to be a missile crewman.

Mitch got a bachelor's in civil engineering at the Colorado School of Mines. He was married while still in college, and during that time helped his father-in-law on a family ranch out near Limon. The father-in-law ran about two hundred Angus and Hereford, and put about two thousand acres in wheat, so Mitch had modestly affluent prospects as a son-in-law rancher. But he didn't want to be stuck forever in a small town. After his graduation and his divorce, he worked in construction jobs, then volunteered for the Air Force as a way out. His first choice for duty was an assignment in civil engineering. His second choice was the Space Command. He also put a check mark in a box indicating that, yes, he would volunteer for missiles if they wanted him. Pilot training would have been nice, Mitch says, but like so many of the others he has bad eyesight. And the Air Force did want him for missiles. He could be an officer, he was told, if he accepted missile-crew training.

The temperature has stalled at around freezing, so the road is now slick as a salamander. We creep along at fifteen miles per hour. We pass the By-Way Cafe, this time without stopping. Some of the other off-duty crews are already snowbound, resigned to staying out overnight at the capsule sites and trying for Great Falls tomorrow. Voices on the two-way car radio can be heard saying things such as, "We are ten-seven here at the By-Way," meaning they are stopped indefinitely, holed up, hunkered down. "Roger, we are not going anywhere. Tell the commander we've destroyed our codes." Our van falls in behind a county snowplow and we follow it slowly up a long hill. We pass a Coors beer truck, parked and abandoned on the shoulder. Like any faithful son of Colorado, Lieutenant Mitchell is known to be partial toward Coors. "Mitch, quick," says his partner, Lieutenant Calkins. "There's some cold ones."

Mitch, the quiet country boy, vouchsafes a sphinxlike smile.

WHEN THE steaks had been eaten and the plates pushed aside, I asked that table full of colonels another question. But before asking it, I warned them, I had to describe something called Newcomb's Paradox. These colonels were polite men, impressively open-minded, and they heard me out.

Newcomb's Paradox is a hypothetical experiment with some interesting implications about logical strategy and human motivation, I said. Possibly even some implications about nuclear deterrence. The experiment involves two boxes, two sums of money, and two human participants. Call one participant the Experimenter; the other is you. The Experimenter presents two boxes. One is transparent, the other black and mysterious. He puts a thousand dollars into the transparent box, and you can see it sitting there. Into the black box, beyond your ken, he will put either a *million* dollars or nothing at all. Then you will be offered a choice: You can take the contents of the black box or the contents of *both* boxes. Depending on how you choose, and on one other catchy factor, you will receive either a million dollars (black box, good luck), or nothing at all (black box, bad luck), or a thousand dollars (both boxes, bad luck), or $1,001,000 (both boxes, good luck). Now here's the catch: The Experimenter will decide whether to put a million dollars into the black box, or no money at all, on the basis of his prediction of your behavior.

If he predicts that you will be greedy and take both boxes, the Experimenter will put nothing in the black box. If that prediction by him is correct (and you know, by the way, from the history of past experiments that he has often predicted correctly), you will get just a thousand dollars. But if he predicts that you will take only the black box, then he *will* put the million dollars in that box. So, I asked the colonels, which box would you take? Which strategy makes the best sense?

Opinion among the colonels was split. Opinion among philosophers is also split. But the man who first published Newcomb's Paradox, the eminent Harvard philosopher Robert Nozick, has argued quite persuasively that the most sensible course is to take both boxes.

Won't that make you likely to get an empty black box? No, says Nozick, it will not—because the Experimenter has already made his prediction before the moment you choose, and *causality as we understand it does not flow from the present to the past.* Now, I said to the colonels, let's talk about American strategic deterrence.

First, grant the premise that the countervailing balance of nuclear weaponry controlled by Moscow and Washington is precisely what has prevented that weaponry—so far—from being used. Grant the premise that nuclear deterrence, ever since Nagasaki, has worked. Maybe that's true and maybe not—but grant it regardless as a premise.

Secondly, imagine this:

You are a missile crewman, seated in a launch-control capsule. The ultimate message arrives. You confirm the validity of the EWO codes, roger. You are in a war environment, roger. Your missiles have been transferred to the *enable* mode; they are launchable. Now the command center back at Malmstrom goes off the air, evidently vaporized by a hit. Malmstrom itself is gone. Great Falls, Montana, is gone. Washington and New York and the rest of the civilized world may or may not be gone. But the president of what at least until recently was the United States is still out there, somewhere, in his last-ditch airborne headquarters, and he sends you the message: "Son, it's time to launch. Turn the key." Deterrence has failed, so he tells you to turn the key.

Now wait. Throughout more than three decades, we have all listened to the righteous claim that these American missiles exist solely for deterrence. America is not interested in a preemptive first strike, says this claim; America is not interested in gratuitous slaughter or revenge; America builds missiles and warheads only for the sake of deterrence. All right, grant that one too.

But if deterrence has already failed, in our hypothetical case, then what use could there be in turning the key? Your own behavior can't undo what has already been done. Your own act of launching missiles can't restore validity to a threat that didn't achieve its purpose. Think of Newcomb's Paradox. Recall that causality as we understand it does not flow from the present to the past.

So I proposed, anyway, over the remains of dinner. And the colonels seemed to ponder it all carefully. They were interested, they were polite; they disagreed.

Two days later I tried the same question on Captain John Cotter. By now he had finished his alert and returned from Oscar-1 through the quiet aftermath of the blizzard. He was eager to have a little time with his family. But we left his wife and his two-year-old daughter at the house and went for a walk in the neighborhood. It was a cold, clear afternoon. We circled through a Malmstrom suburb, ignoring the unshoveled sidewalks and filling our shoes with snow. I described Newcomb's Paradox, then again made the analogy to a missile crewman's position. If deterrence has failed, I asked, what value in turning the key?

With his sweet and disarming directness, Cotter said: "Boy, that's a tough one, yeah."

I added a stipulation that hadn't come up with the colonels. Say that you, John Cotter, are on alert in the Oscar-1 capsule and that your missiles' guidance systems have recently been reprogrammed. American missile crewmen generally don't know where their own missiles are aimed, true enough, but they can't help but be aware of the possibilities. Say that your missiles, this particular day, are not targeted against Soviet silos. They are not even targeted against a young Soviet captain on missile-launch alert in Siberia. They are targeted, rather, against the wife and two-year-old daughter of that young Soviet captain. Deterrence has failed, mushroom clouds have begun blossoming, and you are ordered to launch. What value?

Cotter said: "That's a cause of heartache. I think about that a lot."

Then Cotter took a breath and said: "This job isn't exciting. It isn't a power trip. It's just something that needs to be done. And maybe in ten years, who knows, they'll pull those things out of the ground and they won't be necessary any more. But until then, I just think that it's important that there's somebody who is ready, and committed in his heart, to take the required action."

STOPPING along the road's shoulder is now impossible. There *is* no shoulder—only an icy driving lane, then wind-piled snow, then the ditch. Even to slow down is dangerous. The commercial truckers are still cannonballing along like maniacs, seated high up above the ground-blizzard conditions, heavy with traction and momentum and able to see where they're going, though totally unable to swerve suddenly or come to a quick halt. One of those semis could crush us like a beetle. Most other vehicles have pulled off to wait. We continue. At a turnout in the town of Geyser, we park long enough to break ice off the windshield wipers. Then again onward.

Lieutenant Mitchell is seated beside me. He stares out the window. Mitchell is self-contained and laconic, quite different from John Cotter; intuition tells me not to accost him, at least not yet, with probing personal questions and descriptions of Newcomb's Paradox. Instead we talk about what's outside the car window.

"I see there's some wheat coming up here," Mitchell says at one point. Yes, there's a field of green shoots, now being buried again beneath the new snow. We discuss winter wheat, and Mitchell tells me about wheat mosaic, a disastrous empty-kernel condition that can develop when sprouting winter wheat has been subjected to a hard, late frost. We discuss scours, a disease that can kill newborn calves if they get chilled and weakened from lying in wet snow. I've had neighbors driven nearly bankrupt, and to exhausted despair, trying to rescue their calves from scours. Mitch knows all about scours, and about wheat mosaic, from his labors in Limon, Colorado.

"Farming," he says. "What a risky business."

We ride on toward Great Falls through this ridiculous blizzard. Neither of us is driving, but that's no excuse.

# *Karl's Sense of Snow*

Karl Birkeland, a mild-mannered young man whose sharp features and dark-rimmed glasses give him an uncanny resemblance to Clark Kent, likes to refer to himself as a snow nerd. What he means by that gently self-mocking phrase is two things. First, his professional duties entail immersing himself in the arcana of snow science—that is, crystal formation, temperature-gradient metamorphism, slab analysis, fracture mechanics, and suchlike. Second, he loves his work boundlessly. Karl is an avalanche forecaster attached to the Gallatin National Forest Avalanche Center in Bozeman, Montana. His job is to make daily assessments of snowpack stability and to issue daily advisories that help keep snow-loving recreationists from getting themselves killed. Nerdy like a fox, he has found a professional niche requiring him to spend much of his working time on skis in the backcountry.

This particularly day, Karl stands in a snow pit at 8,500 feet near the west edge of the Yellowstone plateau, not far below Lionhead Peak. It's a fine sunny morning in early March, with a good chance that the afternoon temperature might soar to twenty. The pit, freshly dug and exposing a 220-centimeter column of snowpack, is deeper than Karl is tall. Working beside him is his partner, Ron Johnson, another self-described snow nerd with an unnerdly devotion to backcountry skiing. Using a stem thermometer and a collapsible measuring stick, Ron has been taking depth and temperature readings from various strata in the snowpack, dictating the numbers into Karl's notebook. Now their attention narrows to a single stratum.

"Well, well, well," says Karl, staring at crystals through his Pentax twenty-power monocular. "If it isn't our little friend."

The friend he's referring to, in that discerning and ironic tone, is a dubious one: a thin layer of rimed surface hoar lying forty-five centimeters below the surface, just above a firm icy crust. Something sounds odd, I concede to you, in the mention of "surface hoar" a foot and a half below the surface. But the fact is that surface hoar doesn't necessarily change either in name or in nature as it changes its relative position. At one time it was on the surface, yes, a layer of feathery crystals glittering like mica under the winter sun; but now it's coated with rime and submerged beneath three weeks' worth of later snow. Still, its unique crystalline structure as surface hoar continues to dictate its mechanical properties. And those properties, combined with its present position, carry some sobering implications.

If Karl and Ron were a pair of homicide detectives, this would be their first major clue. In effect it is, and they are.

Karl passes me the monocular and the sample of surface hoar, sprinkled out on a plastic card. Do I see it? he asks. Um, I'm not sure that I do. What I see is a jumble of tiny shapes, some angular, some lumpish, a few beginning to liquefy from the heat of my breath, like a scree pile at a moment of hellish meltdown.

"You have to have looked at a lot of crystals before you start to recognize this stuff," Karl says.

The thing that's dangerous about surface-hoar crystals, he adds, is that "they're really strong in compression and really weak in shear." What he means is that, under pressure from an overburden of snow, they keep their vertical rigidity while developing a hair-trigger susceptibility to horizontal shift. They won't squash, but they may well fracture sideways, toppling like a row of dominoes. When a layer of buried surface hoar does fracture, a sudden "whoomph" sound and a slight cake-fall of the snow surface is one possible consequence. Another is that a slab avalanche cuts loose, carrying a skier away to oblivion or roaring down to entomb a snowmobiler. The cake-fall effect is more likely on a very gentle slope. With steepness greater than thirty degrees, the avalanche becomes more likely.

Grasping these counterintuitive points—surface hoar buried beneath the surface, rigid but shifty, seeming innocent under most circumstances but proving deadly in some—is your first step into the intricacies of snow science. Never mind for the moment just how hoar is created, or how it gets rimed. The general truth is that snow is amazingly complex; the wealth of particulars can't be crammed into a few hasty sentences.

Many of those particulars can't be put into sentences at all. They can only be apprehended, so it seems, along other channels.

OF COURSE the diversity of individual snowflakes is legendary, as reflected in the old saw that no two are alike. But the anatomy and dynamics of aggregate snow, fallen snow piled up by the centimeter, by the foot, by the ton, snow as it settles and transmogrifies within a seasonal snowpack, are even more tricky—and more consequential.

Avalanches kill about fourteen people per year in the U.S. and roughly half as many in Canada, with most victims caught skiing or snowmobiling in the mountains. The fatality toll has risen steeply over the past several decades, reflecting growth in the popularity of winter recreation but no parallel growth in public awareness of the risks. Most backcountry skiers and snowmobilers still go charging out, eager and blithe, with almost no general knowledge about avalanche danger and how to avoid it. The public-education work of Karl Birkeland and Ron Johnson—they speak to school classes and clubs, they conduct workshops at the local university—is intended to cope with that deficit. So are books such as *The Avalanche Handbook,* by David McClung and Peter Schaerer, and *Snow Sense,* by Jill Fredston and Doug Fesler. But general knowledge isn't enough; there's also a need for specific and up-to-date information about local conditions. Given the diversity of landscape in an area like the Gallatin National Forest, an expert assessment of current strengths and stresses within the snowpack can make the difference between an avalanche advisory that diverts people away from the most perilous slopes for a critical few days and a search-and-rescue mission to dig out a body.

Gawking at crystals under a hand lens is only part of it. Scientific scholarship (in meteorology, physics, statistics, a bit of geology and chemistry) is another part. And beyond all the visual and intellectual data stands still another dimension of knowledge: the intuitive savvy that comes only from field experience. Some of that intuitive savvy filters up, quite literally, from below.

"It takes a long time to develop the feel for snow," Ron says. "What Karl calls 'patroller's feet.' It takes years."

Karl himself started skiing at age two and was a junior ski patroller by sixteen, at Lake Eldora, Colorado. He eventually did eight seasons as a patroller, performing avalanche control and other chores at Lake Eldora and elsewhere, in the course of which he spent many thousands of hours attuned to the whisper and crunch of snow underfoot. Ron too served his apprenticeship as a ski patroller. Now they both have master's degrees in snow science, and Karl is at work (time allowing, in the off season) on a doctorate. These two attributes they share, Ron tells me, represent the fundamental requisites for an avalanche forecaster: a nerd brain and patroller's feet.

He understates the demands of his job. It also helps, as I've seen, to have a runner's legs and a mountaineer's lungs. Put all that together with a jeweler's eye for crystal morphology, and you begin to possess what might be called a sense of snow.

IN ONE of the Greenlandic Inuit dialects, the word *qanik* denotes a particular type of snow: big fluffy flakes that fall almost weightlessly and cover the ground like a layer of pulverized frost. My source for this stray fact is a novel titled *Smilla's Sense of Snow,* by a Danish author named Peter Høeg, which I happened to read shortly before my outing with Karl and Ron. The novel's main character, Smilla Jasperson, is a Greenlandic woman, transplanted uncomfortably to Copenhagen, who finds herself drawn into a murder mystery by the sight of footprints in snow on a roof. Smilla is a fiercely detached person, an outsider to Danish society and to human concourse generally; by her own account, she thinks "more highly of snow and ice than love." The telltale snow on the roof is not *qanik* but a very different sort, and the prints fill her with dark suspicions, which she can't (or won't) articulate verbally. What Smilla sees in snow is far more than what the Copenhagen police see.

Smilla in Denmark is a token of the disjunction between snow-world cultures, such as the Inuit, and our own fair-weather, industrialized cultures below the permanent frost line. Those cultural differences reflect differences not just of custom but of perception. Eskimo peoples, we've all heard it said, have a rich lexicon of terms for different kinds of snow. For example, take the variants of snow as acted on by wind: *api* refers to fresh snow not yet disturbed; *siqoq* to snow blown along the ground; *upsik* to wind-packed crust. A writer named Ruth Kirk, in a book simply titled *Snow,* reports that there are dozens of other Eskimo words carrying dozens of other nuances, including "fluffy deep snow that fell without wind; wind-packed snow firm enough to walk on; wet snow belonging to springtime; the walk-anywhere crust of an early morning that comes after a cold night," as well as corn snow, sugary snow, snow in long drifts, snow in round drifts, snow in big drifts, snow in small drifts, snow for traveling, snow for igloos, and more. Presumably there's also a word for the snow that makes good snowballs, and another (or maybe it's the same) for the snow that clumps maddeningly to the bottoms of your touring skis when you pass from shade into sun.

Conventional wisdom suggests that we temperate-zone folk are obtuse by comparison, having a pitiful dearth of words for, and sensibility about, snow. But conventional wisdom, as usual, does no justice to the complicated reality. Conventional wisdom ignores the Russian meteorologist Shuchukevich, who in 1910 reported seeing 246 different kinds of snow crystals during a single year of observations near St. Petersburg. Conventional wisdom neglects the wonderfully demented Wilson A. Bentley, of Jericho, Vermont, who in 1931 published a photographic portfolio consisting of 2,453 separate images of snow crystals. Bentley didn't assign labels to these different crystals, but he did give each one its iota of durable visual fame, like 2,453 weird-looking kids gazing up off the pages of a 1931 high school yearbook.

Conventional wisdom also overlooks the fact that, during

more recent decades, taxonomically inclined snow scientists in the temperate nations have been as busy as squirrels, creating classification systems for the multifariousness of snow. In 1949 the International Commission on Snow and Ice offered a system using letter-and-number codes, and a revised version published in 1957 applied that scheme to ten essential forms: stellar crystals, needles, plates, columns, capped columns, spatial dendrites (whatever those are), hail, graupel, sleet, and a catchall category for the rest, considered irregulars. In 1966 a pair of zealous researchers named Choji Magono and Chung Woo Lee, at Hokkaido University, proposed a more intricate system comprising eighty forms, including some that look more like space junk left in orbit by NASA than like any snowflake a child might imagine at Christmas. Magono and Lee listed needles, bundled needles, sheaths, bundled sheaths, scrolls, cups, pyramids, hollow bullets, solid bullets, stellar crystals with spatial plates, plates with spatial dendrites, dendritic crystals with sector-like ends, hexagonal graupel, cone graupel, rimed needles, rimed columns, and let's not worry about what-all else. The thing you and I need to bear in our snow-swirled minds at this point is that all of these categories apply only to falling and freshly fallen snow. As the stuff piles up on the ground, it becomes subject to other phenomena, such as the formation of surface hoar and the transformation of crystals within the snowpack.

Surface hoar comes into being, somewhat like dew, by the crystallization of water vapor on a clear, humid night when the snow surface is much colder than the air just above it. (Rimed surface hoar, such as that menacing "little friend" that Karl noticed lurking forty-five centimeters beneath the surface at Lionhead, consists of surface-hoar crystals clotted with additional ice that has been added not from vapor but from tiny, supercooled liquid droplets. It's an unusual combination, even to the experienced eyes of a snow nerd; possibly, in this case, it resulted from a very cold fog.) The transformation of buried crystals occurs by at least three different processes, the most notable of which is temperature-gradient metamorphism. This process is driven by the diffusion of vapor from relatively warm (and therefore vapor-laden) layers in the snowpack toward relatively cold (and therefore vapor-lacking) layers. The temperature gradient in most cases runs from deep, warmer layers up toward shallow, colder ones. If that gradient is extreme enough and other conditions are right, the rising vapor causes a metamorphosis of snow crystals into delicately faceted grains, known as depth hoar. A layer of depth hoar, like a layer of buried surface hoar, is something that avalanche forecasters view with attentive distrust. They know that when a great slab cuts loose, a weak layer of hoar may well be the site of the fracture.

But wait, here's another counterintuitive point to be clarified. Did I say that depth versus shallowness, within a snowpack, correlates with warm versus cold? It's true. The ground underneath a deep snowpack remains at thirty-two degrees Fahrenheit all winter, while the topside temperature goes to thirty or forty below, and the various snow strata gradate between those extremes. How so? Because the ground continues radiating heat during winter (partly from stored summer sunshine, partly from Earth's own molten core), whereas the air turns quickly cold and the snowpack itself provides insulation between. That's why you burrow into a snow cave when you're camped in the high country during January, rather than climbing a tree.

This much is only a scratch on the subject of how piled snow changes over time and how those changes affect avalanche potential. Besides temperature-gradient metamorphism, there's also equitemperature metamorphism (which tends to produce rounded grains, strongly bonded layers, and therefore the huge cohesive slabs that make slab avalanches so cataclysmic) and melt-freeze metamorphism (which tends to produce corn snow and therefore great skiing). But not every weak layer is destined to fail and not every strong layer is destined to go sliding away as a murderous slab. A passel of other variables must be considered—incline of the slope, its directional aspect, recent weather history in the area, and more. No wonder you need a master's degree, just for starters, to grasp what's happening under a tilted white meadow.

KARL BIRKELAND'S master's degree project gave him a hard-won appreciation for the difficulty of predicting the behavior of snow. His basic question was: How much does the strength of a snowpack vary, over short distances, as a result of minor differences in substrate, exposure, and snow depth? He took hundreds of measurements in two different sites during two consecutive winters, using a fancy tool called a digital resistograph, and succeeded in proving . . . what? Well, to simplify grossly, he succeeded in proving that snow is even more complicated than digitalized snow science. You can mimic it with a theoretical model, but not precisely; you can predict its behavior from a few measurable parameters, but not accurately; you can generalize from one site to another, but at peril of being wildly wrong. This didn't surprise Karl and it shouldn't surprise us, since snow is a form of water, after all, and water's dynamic complexity is virtually infinite.

Snow truths do exist, but they're elusive and protean. Like snowflakes themselves, they tend to melt away when carried indoors. That's why professional avalanche forecasters with real-world responsibilities, like Karl and Ron, spend such a large portion of their time in the field, looking at the snow, listening to the snow, touching the snow, reconnecting with this mysterious substance they think they know. One mode of reconnection is by performing quick-and-dirty experiments, known as stability tests, on as many different snowy hillsides as possible. These stability tests provide rough answers to basic questions. Will a layer of depth hoar or buried surface hoar give way under disturbance? If so, which layer, at what depth? What degree of disturbance will be necessary? How weak is the weakest layer? A reasonably good stability test—the Rutschblock test, for instance, developed in Switzerland two decades ago—gives answers that can be codified numerically, thereby allowing crisp comparisons between one situation and another. But since even the Rutschblock test has its shortcomings for practical forecasting, Ron and Karl have developed a test of their own. They call it the stuffblock test. Unlike virtually all other stability tests, theirs is both convenient and precise.

Having finished gathering their temperature and depth readings in the snow pit, they proceed proudly to give me a demonstration. Using a snow saw, Ron carves out a square-sided pillar, leaving it freestanding at its full 220-centimeter height. From his ski pack Karl produces a Zebco DeLiar fish scale (model 228, designed for lunkers up to about forty pounds) and a plain nylon stuff sack (the type used for holding a sleeping bag), from the bottom of which dangles a calibrated string. Herein lies the convenience of the method: Both items are easily portable and their use will take only a short time. Karl fills the sack with snow—exactly ten pounds of snow, as gauged by the Zebco DeLiar. Ron takes the sack and sets it gingerly atop the snow column. Nothing much happens. This itself is a datum. Under pressure from the standardized weight, the column has held.

Now Ron raises the sack exactly ten centimeters, as gauged by the dangling string. Herein lies the method's precision: The standardized weight is dropped from a distance that's variable but easily measured. Ron lets the sack drop. The snow pillar cracks away at its weakest stratum—the rimed surface hoar forty-five centimeters below the surface.

Repeating their stuffblock procedure on several more pillars, Ron and Karl find some slight variation. The second pillar cracks away—again, forty-five centimeters down—under only the gently applied ten-pound weight. Likewise with the third pillar. To me it's all just crumbling snow, but to Karl and Ron these tests yield a stark conclusion. The snowpack in which we're standing is very damn unstable.

THE REAL scope of that instability becomes clear just a few minutes later. As Ron packs away his snow tools, Karl skis back across the brow of the slope, moving out beyond a small cluster of fir trees that shield our snow pit from the broader sweep of mountainside. Keenly aware of the danger, he stays just high enough to avoid the precarious zone of incline. Suddenly we hear him hoot. Turning, Ron and I see a roiling crest of whiteness cascading downhill, just a few yards beyond and below where Karl stands. It's a slab avalanche in progress, presumably triggered by him.

The whole slope, except for our tree-anchored part of it, has fractured and run. All the neat crystal laminations have collapsed into chaos. The traveling front stretches eighty yards wide, encompassing God knows how many tons of snow. I've never seen such a large avalanche at such close range. It foams. It smokes. It moves like a wave crashing onto the beach at Waimea Bay. It's scary, mesmeric, quite big enough to have snatched all three of us down in a homicidal hug, and almost utterly silent. It tears into the trees below, culminating like the last howling chords of a Beethoven symphony, and then it sends up a high exhalation of powder.

It's over. The sun is still shining. Nobody's dead.

Afloat in a rhapsody of appreciation, Karl stands on the unbroken rim of the slope, expressing himself in a series of fervent exclamations. The gist of these exclamations pertains to the terrible, forceful, paradoxical beauty of snow. The gist is that this avalanche forecaster loves a good avalanche. Ready with my snow-soggy little notebook, I wait vulturously for him to say something quotable. But he doesn't. He doesn't. Instead, just a few earnest yelps and hoorahs. It's an extraordinary moment, this direct transfer of energy from Karl to the avalanche and back to Karl, but at the verbal level, nothing remarkable. At one point he simply hollers, "All *right!*" We've come to the point where words, like numbers, are inadequate representations.

Karl himself understands the inadequacy as well as anyone. Several days later, by telephone, we discuss it. He quotes me a passage in a recent novel that, he feels, puts the point nicely: "Reading snow is like listening to music. To describe what you've read is like explaining music in writing." The novel is *Smilla's Sense of Snow.*

"Your nerd brain is, like, always trying to explain snow to other people," Karl says, with some hint of frustration at the enterprise. "Your patroller's feet are how you hear the music." His sense is that it's glorious sonic terrain.

# *The Trees Cry Out on Currawong Moor*

The man's real name, his birth name, is lost in the mists of northwestern Tasmania. He seems to have left it behind, irretrievably, when he was brought into the cold comforts of civilization. Probably it would seem a tongue-twister to us, as do other Tasmanian Aboriginal names—Toeernac, Wobberrertee, Calerwarrermeer, Drummernerloonner. His mother was called Nabrunga. What she called him, we'll never know. His adopted name, given him as a child among the sanctimonious white jailers, was William Lanney.

He became a conspicuous public figure in mid-nineteenth-century Tasmania, this William Lanney. Years later, during his lonely adulthood around the city of Hobart, some of the whites dubbed him "King Billy." It was mockery, but they probably let themselves think it contained a strain of affection. He was a big, boyish fellow with black skin, curly hair, and an expression, at least for the daguerreotype camera, of sad and long-suffering calm. His shoulders were round and his eyes were liquidy. They called him the last living male of the Tasmanian Aboriginal race.

That was misleading, as we'll see, but convenient. They called him more than one thing that he wasn't. Certain dry-hearted pranksters even went so far as to introduce him to visiting British royalty as "King of the Tasmanians." Jesus of Nazareth was once labeled "King of the Jews" in roughly the same droll spirit.

Lanney had been born into a different language, a different world, somewhere along the northwestern coast of the island. In the last days before their capture, he and his family had moved inland and upland, finding their way among the glacier-carved peaks, the eucalyptus forests, the high foggy moorlands near Cradle Mountain. A century and a half later, I've come to search for him here among those same peaks, forests, moors. The polite phrase for such a presumptuous enterprise is *fool's errand*. At very least, though, I'll get exercise and scenery.

WITHIN A few days after landing in Hobart, the capital, I've learned this much: Tasmania is complicated. It's an oxymoronic place, combining the gentility of English-style tearooms with the raw sort of high-country wilderness I know from Montana; a former penal settlement, notorious for its harshness, graced today by an especially free-spirited variant of the breezy Australian charm; a wet, chilly landscape that supports its own version of temperate rainforest on the far side of the tropics. Profoundly alien, yet somehow oddly familiar, Tasmania defies categories and confutes expectations. It's part of Australia (the smallest state) yet detached, a separate island down under the Down Under. It's a final refuge for some of the rarest marsupial species. Even the Tasmanian Aborigines are distinct, having been isolated genetically and culturally from mainland Aborigines throughout most of the past twelve thousand years. In fact, some observers say that the Tasmanian Aborigines were, until Europeans arrived, the most isolated group of humans on Earth.

When that isolation was suddenly breached—by explorers, by whalers and sealers, then by British penal authorities and convicts and eventually by settlers—the very distinctness of the Tasmanians became their great disadvantage. It left them vulnerable to disease and conquest. It led to talk of "domestication," "hybrids," and finally "extinction," as though they themselves were an endemic marsupial species. It made possible all manner of perniciously simplified notions, including the one that William Lanney was the last Tasmanian Aboriginal man.

I have come to this area because I'm interested in the factual and emotional realities of Lanney's life. I plan to walk in the mountains for five days and meditate over a ghost. I want to glimpse his landscape and try to imagine a few of his feelings. But as I begin my little trek, I don't know so much as his name—not even Lanney, the adopted one. I know only that he or someone like him once made a last fearful escape along this same mountain trail. I have a map and a three-sentence clue to guide me.

The map is a pocket-size topographical projection of Cradle Mountain/Lake St. Clair National Park. The three-sentence clue appears on a display panel at the visitors' center near the north end of the park. "After the 1835 drive by George Augustus Robinson to collect Aborigines and establish them on a mission at Flinders Island, the Cradle Mountain area became a refuge for one Aboriginal family," it begins. I've read elsewhere about Robinson—another shimmering, ambivalent figure in Tasmanian history, famed for his "friendly" crusade to round up the native people and stuff them away on an offshore reservation, ostensibly for their own protection. By 1836 he believed that he'd gotten them all, except for the stubbornly elusive Cradle Mountain family. "These people refused to join the mission and were probably the last Tasmanian Aborigines to follow a traditional lifestyle. The family was finally captured in 1842." During a brief stop at the visitors' center, I copy those three sentences into my notebook, not yet aware of the human particulars, the gruesome climax to this story, or how often the words *last* and *finally* will recur.

In addition to the map and the notebook, I'm carrying a kilogram of rice, a kilogram of muesli, sixty dollars' worth of dried fruit, a few freeze-dried dinners, a tent, a rain parka, rain pants, cold-weather clothing, binoculars, a head lamp, a bird book, a mammal book, a pint of Jim Beam, a detailed guidebook on Australian backcountry hikes (bushwalks, the Aussies call them), and a paperback history of Tasmania. Before even reaching the trailhead, I realize it's too much. My pack is bloated and leaden. So I rip six crucial pages out of the bushwalking guide, stow them in a pocket, and give the book away. Just before dawn, on Wednesday of Easter week, I start walking.

The trail, which will lead me upward past Cradle Mountain

and across the moors and on down to Lake St. Clair along an old Aboriginal route, is famous among Australian bushwalkers as the Overland Track. The map tells me I have eighty-four kilometers to cover by Sunday. The air is cool with antipodal autumn.

THE CONFLICT between Tasmanian Aborigines and British colonists broke into full flame during the late 1820s. Earlier there had been only growing tension—due to competition for kangaroo meat, competition for land, the kidnapping of Aboriginal women by horny seal-hunters, and the slaughter of sheep by Aborigines who saw these insatiable beasts displacing their own game. That tension had erupted intermittently in the form of murderous raids committed by one side or the other. Beginning in 1828, it was war. Of course the Aborigines couldn't win. Their weapons were simple, they were outnumbered, and the landscape upon which their survival depended was being transformed. But they were a formidable guerrilla force, and they killed their share of the enemy. So in 1830, the British governor of the colony mounted an operation known as the Black Line.

The Black Line wasn't the most bloody episode in the annals of colonial conquest, or even in Tasmanian history, but it merits its own special infamy as a symbol of British determination to sweep the Tasmanian Aboriginal people off the map. A human line was assembled, comprising two thousand soldiers and convicts and colonists. They were armed with a thousand firearms, thirty thousand rounds of ammunition, and (optimistically) three hundred pairs of handcuffs. Over the course of three weeks, the line moved across the settled region of the island, thrashing along through the woods and meadows to drive a concealed enemy before it. The idea was that this marching cordon would push the last few hundred Aborigines into a small peninsula on the southeast coast, where they would be permanently confined. One man and a boy were captured; two others, shot. The rest of the Aborigines of the region slipped between or around the drivers and away into safer parts of the island.

So something else had to be tried. The chief colonial official appointed George Augustus Robinson to travel across Tasmania, contact the remaining Aborigines, and persuade them to trade their liberty for a form of wardship. Robinson could offer them clothes, food, amnesty, education, Christianity; he would eventually also deliver anomie, malnutrition, despair, and an epidemic rate of mortality. This dubious bargain was euphemistically called "conciliation," and Robinson became known as The Conciliator, a label almost as misleading as King Billy.

THE FIRST half-day's terrain is the steepest. Just a few kilometers south of my starting point, the trail rises straight up at the blind end of a lake and onto the Cradle Plateau. At some points it's a near-vertical ladder of tree roots—only passable, I find, on all fours. The trail rises again up the lower slope of Cradle Mountain itself, then levels into a high track traversing southwestward amid scrub and scree and stands of dead eucalyptus trees, just under the scowl of the peak. But the peak shows its scowl only in flashes, remaining mostly concealed by fog. The fog banks slide open and then close again with surprising quickness, like lumbering aluminum doors to a huge, mysterious hangar. Under these melodramatic conditions, Cradle Mountain resembles a Turner seascape of Gibraltar. Presumably it was magical for the Aborigines too. But we know almost nothing about their beliefs.

I stop for lunch on the plateau beside a small tarn. Quickly my sweat goes cold, and the rain begins. I've been warned about the weather. In this part of Tasmania at this time of year, drizzle and fog are routine, cold rain is likely, overnight snows aren't unusual, and sunshine should be considered an unwonted grace. The park's cautionary advisements are strident about hypothermia. Evidently there have been more people lost to exposure than to falls. I pull out the rain pants and parka. I cover myself and my pack, nap briefly, then hike on.

The traditional Tasmanian men went naked. The women sometimes wore a kangaroo pelt over the shoulders, less for their own protection than to carry an infant. Unlike the mainland Aborigines, the Tasmanians had no sewing techniques. According to what I've read—and I remember it unenvyingly now, as I sludge along—they kept themselves warm with a layer of grease. Given the stern climate, their barefoot and grease-cloaked travails seem almost unimaginable; but they survived for more than twelve thousand years. How? Partly, no doubt, by avoiding unnecessary exposure. This high boggy plateau, for instance, wind-raked and stark, appears bereft of animal prey and shelter. Why would an Aboriginal family leave the more hospitable lowlands, as William Lanney's did, and venture up here? Possibly because they were following a shortcut between milder areas. Or possibly because they knew someone was chasing them.

Late in the afternoon, exhausted, muddy to the shins and under a steady drizzle, I find myself crossing a stretch of moorland that slopes off into fog along both sides. I'm on a ridge. The fog is thick, the light is yellowish gray, and visibility doesn't extend fifty meters in any direction. I can see the foot trail before me, ripped down through buttongrass by previous footsteps. I can see some low scrubby trees scattered sparsely roundabout. I could be on a Hollywood set concocted for one of those B-movie adaptations of Poe, an impression reinforced when, in a moment, I find myself surrounded by large, black birds that look roughly like ravens.

Their eyes, unlike raven eyes, are a creepy shade of yellow. Daubs of white amid their dark wing and tail plumage flash like subliminal parentheses as they fly. My bird book identifies them as *Strepera fuliginosa*, the black currawong, a species found only in Tasmania. One by one they dive at me, passing low overhead, making ratchety calls. I stop. I gape. A currawong swoops close, pulls himself up, and lands atop a stunted tree. He sings his hoarse croak. I croak back.

Raucous and bold and so agitated by my intrusion, the birds give this desolate place a weird charm. Already in my brain it has become Currawong Moor. I linger. They dive, they feint at me, they withdraw. They sit on their snaggy perches and then, following as I move on, they dive again. Their noisy lunges are not quite attacks. A biologist would call it mobbing behavior. But why here, why now, and why at me? Are there unfledged young that need protecting in nearby nests? Not likely, at this time of year. Is it a generalized sort of territorial defense? Or are the birds delivering, maybe, some other message? Do they scream for themselves, these bile-eyed creatures, or are they proxies?

Passingly I consider the notion that here on Currawong Moor reside the spirits of those last unconciliated Tasmanian Aborigines. The place does have a preternatural feel. But I'm no mystic. No, I decide. No, that one doesn't fly.

I walk on. I stumble down off the ridge to a sheltered campsite. Later I learn that the gnarled trees along that foggy stretch of moorland, favored as perches by the currawongs, are *Athrotaxis selaginoides,* another species unique to Tasmania. The common name is King Billy pine.

BY 1835, George Augustus Robinson had persuaded about three hundred Aborigines to accept deportation to Flinders Island, a bleak lump of land off the northeast coast. Many died in transit camps. Among those who reached the compound on Flinders, many more died soon, notably from respiratory infections such as influenza and pneumonia. Their daily rations consisted mainly of salted meat, flour, sugar, and tea. The rate of infant mortality was high. Within a few years, the three hundred original captives had been winnowed to about a hundred survivors. Of the rest of the Tasmanian Aborigines, most had either been kidnapped by sealers (who took women away to their camps, making them slaves and concubines) or shot. The native population of Tasmania had always been small. Now just a single family was thought to be still at large.

In 1836, a posse led by Robinson's son found that family near Cradle Mountain. To his father's dismay, George Junior let them escape.

They remained fugitive for another six years. Then in 1842—possibly because they feared being shot, possibly because their land and their options had been stolen, possibly from loneliness—they gave themselves up. According to a historian named Lyndall Ryan, author of *The Aboriginal Tasmanians,* the last family "consisted of William Lanney, aged seven, his parents, and four brothers." They were shipped to Flinders. Within five years, both parents and three of the brothers were dead.

William himself had a somewhat hardier constitution. Eventually the remnants of the Flinders population, including him, were moved to another site. Their new compound was an old convict station at Oyster Cove, not far from Hobart. The station was in ruins, windswept, unsanitary, and the rations provided there were marginal. By 1859, just fourteen people survived. One woman was constantly in tears. Another woman spoke only to her dogs. Of the five men, one was senile and almost blind, two others were alcoholics. As reported by Ryan: "The women most admired William Lanney, aged twenty-four, whom they saw as 'a fine young man, plenty beard, plenty laugh, very good, that fellow.' "

He escaped from the dead-end misery of Oyster Cove, going to sea as a whaler.

ON THURSDAY it's more fog, more drizzle, more long stretches of delicate moorland across which the foot traffic has cut a muddy gash. At one point the mud sucks me down to my kneecaps. I struggle carefully to extract not just my feet but also my shoes, a woebegone pair of runners, which are precious, since I was too stupid to bring an extra pair. And now I discover, oho, that I'm in leech territory. Tasmanian moorland leeches, it turns out, are thin little demons that move like inchworms—resembling the rainforest leeches of Madagascar, much daintier than the lake leeches of Wisconsin. A few of these beasts sneak up my trouser legs, drink deeply, and get away clean, leaving their bloody drool. In the evening I find a dead leech that has poisoned itself by eating chemicals off the glossy paper of my map. Serves the sucker right.

Mere leeches can't spoil a good day. I've gotten a glimpse of the great dolerite columns of Mt. Oakleigh, far out across the Forth River gorge. I've hiked through a handsome stand of Tasmanian beech trees, their leaves just going yellow with autumn. And I've found a pretty riverside campsite, a clearing grazed down to smooth lawn by the loving attention of wallabies.

On Friday, which happens to be Good Friday, the early morning air is decidedly cooler. I'm up making coffee and hot muesli by head lamp. But I dally around camp until the sun is clear of the mountains. Then I hike out of the Forth River drainage into another, through a mixed forest dominated by ancient eucalyptus on a trail paved with those yellowed beech leaves.

In the course of the day I see a dozen other hikers. Easter weekend is a getaway time for Tasmanians, last gasp of the summer season, like our Labor Day. Mentally now I'm measuring the remaining kilometers against the remaining time, and figuring my increments of daily distance so I can camp where other people don't. Hiking the Overland Track can be a sociable experience, if you aren't careful.

What I want is solitude. I'm trying to conjure some emotional empathy with that last Aboriginal family. For this, I find the moors preferable to the eucalyptus forest, since the sweeps of moorland are more alien to my own eyes and feet. What did those people feel, up here, during their years of eluding the white man? They knew that the world had changed around them, gone strange and inimical, but they couldn't have comprehended why. They must have been angry, terrified, lonely; certainly they were confused. Closest approximation in our culture, it occurs to me, might be the diary of Anne Frank.

WILLIAM LANNEY never married. So far as history knows, he never fathered any children. He outlived all the other men among the "conciliated" Tasmanians and therefore, toward the end of his short life, he enjoyed an unenviable fame as the last male of his race. He was repeatedly photographed. Some individuals who considered themselves scientists also found him an object of interest. But unlike the photographers, these men had to wait until he was dead.

He died abruptly, in March 1869, at a Hobart hotel. Possibly it was from cholera, though the diagnosis is uncertain and subsequent events didn't help to clarify it. He was thirty-four. His body was sent to the morgue at Colonial Hospital, supposedly to protect it from grave-robbers. One of the leading local grave-robbers, however, happened to be the hospital's chief surgeon, Dr. George Stokell. Another was Dr. W.L. Crowther, George Stokell's avid rival. Having maneuvered Dr. Stokell out of position, Dr. Crowther went to the morgue that night and decapitated William Lanney's body. According to various reports, he either replaced Lanney's head with the head of another dead man or else he took Lanney's skull only, and replaced that with the other man's, wrapping the skin of Lanney's face back around it. Whatever his grisly procedure, Crowther worked fast. At 9:00 P.M. the gatekeeper saw him carry a parcel away under his arm. Soon afterward, Stokell returned. Frustrated at having been preempted, he performed his own pseudoscientific mutilations, cutting off Lanney's hands and feet. It bears mentioning that Stokell wasn't a lone ghoul; after a quick consultation, his acts were endorsed by his colleagues of the Tasmanian Royal Society.

The next afternoon, following an Anglican funeral, William Lanney's much-abused body was buried at St. David's Cemetery. His whaling friends, who knew him as a shipmate not an anthropological curio, saw him into the ground. After nightfall, Stokell—not satisfied with his earlier hackwork—exhumed the corpse. Back at the hospital, he and other members of the Society indulged themselves with a midnight examination of Lanney's remains. Still later, Crowther too came back to the morgue and, crazed by his own thirst for knowledge, broke down the door with an axe. But except for a few bits of flesh, some blood and fat, Lanney's body by that time was gone. Stokell had moved it again. Eventually the mangled remnants of "the last Tasmanian Aboriginal man" went into a cask, and then to another cemetery. Stokell kept enough of the skin to have a tobacco pouch made for himself.

None of this was deemed criminal behavior. Crowther was suspended from the hospital. Stokell, having better political connections, wasn't. At a public inquiry, both men declared their innocence of anything less savory than legitimate science. Lyndall Ryan writes that, besides Stokell with his tobacco-pouch keepsake, "other worthy scientists had possession of the ears, the nose, and a piece of Lanney's arm. The hands and feet were later found in the Royal Society's rooms in Argyle Street, but the head never reappeared."

So it's no wonder that, when the "last" Aboriginal woman was on her deathbed, a few years later, she said, "Don't let them cut me up."

WILLIAM LANNEY was not actually the last Tasmanian Aboriginal male. At most he might be considered the last male of unmixed descent. Does that categorical claim—his ancestry was unmixed, he was the last—carry overriding significance? Some voices in modern Tasmania have said so, arguing that only "fullblood" Aborigines should be counted as truly Aboriginal and that, since none such survive, the indigenous Tasmanian race has ceased to exist. This presumption accords well with continuing governance and ownership of the island by whites.

But in a world where pale-skinned colonialists and slaveholders have so often treated "non-white" as an absolute category, even as applied to persons with only a fraction of dark-skinned ancestry, the notion of inverting the invidious blood standard when that's suddenly more expedient seems ridiculous. Surely a "halfblood" Aborigine who has been raised as an Aborigine and feels like an Aborigine, and who has been treated disadvantageously as a non-white, ought to be entitled to claim to be Aborigine. The real significance of William Lanney, less ambiguously, is that he was the last surviving male of that disconsolate group who had accepted George Augustus Robinson's offer of conciliation.

In the meantime, many of the Aboriginal women kidnapped by white sealers had left offspring. The mixed-blood offspring had generally been raised as the racial and cultural heirs of their mothers, not of their fathers. They hunted muttonbirds on the little islands off the northern coast, they strung shell necklaces, they practiced other folkways and crafts in the tradition of their Tasmanian ancestors. Unlike the Robinson group, they had a birth rate that exceeded their death rate. Eventually they gathered together on an unfertile mound of granite that no one else wanted, a place known tellingly as Cape Barren Island. In 1912, the government of Tasmania granted them rights to that island, as a communal reserve, and in 1951 those same rights were rescinded.

The ostensible reason for abolishing Cape Barren Reserve was that the Tasmanian Aboriginal people were extinct, and that therefore the islanders—whoever or whatever they were—should assimilate themselves into white society. The other reason was that land on Cape Barren Island was now an economically estimable commodity. White people with hard cash coveted it.

The myth of the total extinction of the Tasmanian Aborigines has been carried forward through history books and popular culture by two sorts of momentum. First, it was more dramatic than the reality—and everyone likes drama. Second, it was more convenient. Extinction of a people demands only remorse. Living Aborigines demand land.

Today the government of Tasmania is more enlightened on these matters. But the struggle for land rights and recognition is still being waged by a small group of Aboriginal descendants. Back in Hobart after my bushwalk, I speak with one, a lanky mid-thirtyish woman named Denise Gardner. She works at the Tasmanian Aboriginal Centre, a no-frills operation based in a rickety old house. After we have talked for a half hour about land claims, legal-aid services, substance-abuse programs, and other matters, Gardner tells me that there's one further rankling issue. Her eyes widen fiercely. She mentions the mutilation of William Lanney. She mentions the tobacco pouch.

"Our belief is that Aboriginal people are entitled to a decent burial," she says. "We don't accept our ancestors' remains sitting in museums. We're going to get our ancestors back and set their spirits free." Some progress has been made, she says. The skeleton of the "last" Aboriginal woman, Truganini, displayed for years in a Hobart museum, was returned to the Aboriginal community in 1976. It was cremated. Other remains have come back from museums outside Tasmania. And in 1985, Gardner says somberly, they were given the Crowther Collection. It too was cremated, she says. A private ceremony.

She doesn't specify just what it was, this Crowther Collection, and in a moment of faintheartedness I fail my journalistic duty: I don't ask whether it included the lost head of William Lanney.

COLD WIND, rain, soggy feet. Saturday I hike in a wool hat and gloves. I pass through a chilly rainforest of towering eucalyptus and shade-loving mosses and ferns. In late afternoon I reach the Narcissus River and set up my camp there in the rain. By now I've descended into the southern valley of the park, meeting this cold front as it moves north. I'm lucky that the bad weather hasn't caught me out on those high moors beside Cradle Mountain. I rig a tarp and do my cooking underneath. I crawl into the tent and undress. No leech bites today, I notice. Has it gotten too cold for a leech to care about blood? The season does seem to be turning. Tonight I sleep in wool socks, dry pants, a longjohn shirt, a wool sweater, a down vest, and the wool hat, yet through the deep hours I'm barely comfortable.

Before sleeping I scribble some thoughts. I'm still gnawing on the subject of that one fugitive family. What kept them going for so long? Was it pride? Did they feel a sense of relief, or a sense of shame, when they finally came in off the moors? Or just despair? And then what happened? Did they go to Flinders Island and wither away? My information at this point, recall, is limited to three vague sentences from the visitors' center. I have no names—not even the English-style names branded on as vouchers of conquest. "Who were those last people, at large in Cradle?" I write. "And what became of them?" I make a note to find out. Eventually that resolution will lead me to Lyndall Ryan's book and other sources mentioning William Lanney. But first there's a hike to finish.

I'm awake before dawn, energized by finality, patching my right foot with moleskin for the last day's walk. No drumming of rain on the tent, thank God. I cook breakfast without need to cower under the tarp. I drink coffee and stare at the river. The sky lightens to gray. Then, briefly, there's a large swatch of blue. But the clouds close again.

It's Easter morning. Which means, in Tasmania, that winter will soon be here.

# *The Big Turn*

It's Thursday before the big event and, despite some cause for concern, the impresario is calm. The impresario is a compact 42-year-old man with the grace of a shortstop, the heart of a teacher, the drive of a promoter, and the soul of a clown. His name is Dickie Hall. The event is his 20th Annual Telemark Festival, scheduled for this weekend at the Sugarbush ski area in Vermont.

As many as a thousand telemark skiers are expected. On Saturday there will be a slalom race, on Saturday night a lodge party lubricated by an excellent Vermont-brewed beer, and on Sunday morning a bump contest (that is, mogul-skiing in the freestyle mode, only rather freer even than usual), with various other diversions also happening meanwhile—demo equipment available, daffy costumes encouraged, instructional clinics for all skill levels and ages. "It's like a cross between P.T. Barnum, the Olympics, and a Dead concert," Hall explains, without belaboring the point that he himself fills the role of Barnum. He's a large character but by no means an egotist. Finally on Sunday afternoon folks will converge on the lowermost slope, just above the lodge, for a special symbolic event: a World Record Attempt Group Telemark

Turn, as it's billed in the flyer, with the current record standing at 187. What the record means is that, at a previous festival, 187 giddy free-heel skiers linked hands and started intrepidly downhill, then attempted to execute something loosely resembling a synchronized turn. The best part about it, Hall says, is seeing 187 different ways to fall down.

If his hope is fulfilled, Sunday's big turn will be grander and more giddy than any before. This year's festival *is* the twentieth, after all, harkening back across two decades' worth of good-spirited tomfoolery and athletic dedication to the time when Hall himself was one of the first free-heel ski bums in America. But well-catered tomfoolery for a thousand visiting skiers demands some serious preparation, and on Thursday there's still much to be done.

Hall has been awake since 4:00 A.M., working on last-minute details and devoting a modest bit of concern to the weather. Snow coverage on the mountain is skimpy. The winter has been dry and, notwithstanding the resort's assiduous snow-wranglers, an unseasonal March thaw has stripped the runs half-raw, exposing large marblings of dirt, grass, and rock. Now there's a storm predicted for Friday—but will it hit? If it does, will it contribute anything more than a deceptive skiff over the bad stretches? Should he worry about too little snow, then, or about a few inches too much? It could make a man crazy. The third alternative, which better suits Hall's disposition, is no worrying whatsoever. After all, is this brain surgery at issue here? No. Global peace? No. What it is, merely, is a ski party for pinheads. (The term *pinhead*, as Hall and others use it, is an affably self-derisive double entendre, with only one of those entendres referring to the toe-pin bindings of telemark gear. See "Pinhead Secrets.") Fussbudgetry being therefore inappropriate, Hall has escaped from his office to conduct a site inspection—and, not incidentally, to make a few turns.

I sit beside him on the lift, riding up through a tepid, damp fog. Below us, the slope is punctuated abundantly with knobs of bare granite, smears of mud, and patches of turf, some of the latter looking lush enough to mow. A careful skier could find a path down, it appears, but not without pussyfooting across those gaps. Where I come from in the northern Rockies, by the time a hill has ripened to this stage it has long since been deserted for kayaking or softball. But Hall's reaction is upbeat. Oh, this—this is *nice* snow, he says. Yeah, he can certainly see a party on *this.* Having only just met the man, I mistake his genuine equanimity for forced cheer.

As we near the top tower, the fog turns to a steady wet drizzle. I raise the hood of my parka, hunkering. But the air is warm, smelling freshly of pine duff, and Hall is relaxed. Soaked hair plastered onto his forehead, he gives a small sigh of contentment and says, "I've never had a bad day of skiing in the rain." A bit of rain on corn snow, Hall adds savoringly, is "like butter on silk."

It's my first hint that there's a philosopher hiding inside the teacher who's hiding inside this particular impresario.

DICKIE HALL was a professional skier by the age of seventeen, working as a patroller at certain ski areas out in Colorado and then returning east to do the same in Vermont. He worked the 1973–74 winter at Killington while living out of a tepee. In fact, he was a "tepee hippie" in those days, he says, without specifying the pharmacological details. He got fired from Killington after hijacking the chair lift one night for a larkish moonlight run; trapped halfway up when the bosses cut power to the lift, Dickie and his pals managed to escape with the help of a rescue rope they had foresightedly brought, but in the cold light of following days they still took the blame and the fall.

"That was sort of the start of my nordic career," Dickie recalls. "Because I needed to make some money." He talked his way into a new job as a nordic (that is, cross-country) ski instructor on the basis of quick study and a smooth tongue. Had he actually done any previous nordic instruction? Uh, no, but he'd read a book.

It was the early 1970s. Sideburns were long, pants were bell-bottomed, skiing was still in its late-medieval period, and Telemark was a province in Norway, not an American verb.

Though his experience was in alpine technique, Dickie fell deeply in love with nordic skiing because it allowed him to prowl the whole mountain, any mountain, resort or backcountry, up, down, or sideways. He liked to tour through the woods, cross boundaries, and violate expectations; he liked the sensation of skiing with flexible boots resembling actual human feet. But something was missing from nordic gear and nordic technique as he found it in New England: dexterous control. Then in an old ski book he saw a picture showing the telemark turn, developed a century earlier by the great Norwegian jumper Sondre Norheim. Although the book dismissed it as an archaic maneuver only used for showing off, to Dickie it suggested a way of combining nordic mobility with alpine control. "I realized *that* could be the missing link between the two sports." Next day, with a pair of simple wooden cross-country skis, he tried his first-ever telemark turn. "I got that little swoop, even on my woodies," he remembers vividly. It changed his life.

He began searching for a pair of cross-country skis with metal edges, which would vastly enhance his control. There were none to be had, until he persuaded the Karhu company to send him a pair of its camouflage-white model, made for ski troops of the Finnish army. He matched the Karhus with a used pair of Vasque ski-mountaineering boots, several sizes too big, which he got from a friend. Now he had one set of gear that would work for backcountry yet also allow him to carve turns down the steep slopes of any resort in Vermont. Meanwhile, out in Crested Butte, Colorado, a fellow named Rick Borkovec and some like-minded skiers were also rediscovering the telemark turn, but in his own part of the country, Dickie Hall was a lone pioneer. Did the new approach catch on quickly, I ask, or was he just completely crazy as far as other folks were concerned?

"I was certifiable."

After a year or so, though, he started to win over a few of his more adventurous friends, both from alpine and from nordic, as they saw that he could ski anything they skied, and more. Unaware of the Crested Butte group, Dickie and a handful of cronies became a cadre unto themselves, practicing the happy subversion of telemark and assuming they were the only people on the planet doing it. They even gave themselves an acronymic aegis—BLOPSTI, standing for Blown Out Professional Ski Touring Instructors, a tweak at the actual sanctioning organization (EPSTI, or the Eastern Professional Ski Touring Instructors) from which most of them had emerged. On a nice spring day in 1975, they gathered at the Pico resort near Killington for a party in celebration of telemark skiing, good company, and their own ragamuffin sense of liberty.

"That was the first festival," Dickie says. He still has the group photo.

About a half-dozen years later—years during which Dickie had been teaching telemark, staging telemark races at various New England areas, and promoting this new brand of old-fangled skiing in every other way he could dream up—the notional BLOPSTI aegis was replaced by another, only slightly less notional: the North American Telemark Organization, also known as NATO. The dignified name and the copycat acronym are Dickie's way of flashing a wink at those people who take their skiing, or their world in general, too seriously. "It's not a *real* organization," he reminds me when I ask when exactly it was founded. Real or not, this NATO has thrived and grown. On the efforts of a small paid staff and plenty of volunteers, it now sponsors dozens of telemark clinics, workshops, races, and camps every year, from Tuckerman Ravine to Denali, from West Virginia to Japan. And of course NATO, not Dickie alone, is the sponsor of the 20th Annual Telemark Festival at Sugarbush.

It's a vest-pocket operation with global reach. It has an 800 number. It addresses the public (as its founder too sometimes does, though not often) with a straight face. The best way of avoiding confusion between Dickie Hall's NATO and the NATO of Bill Clinton and Helmut Kohl is to remember this: Dickie's is the one based in a converted loft behind an old wooden house in Waitsfield, Vermont, furnished with a computer, a phone, a telescope, a Japanese kite, a poster of a Viking warrior making a telemark turn, a gumball machine full of M&M's, and a collection of toy trucks. Clinton and Kohl's is the other one.

OVER THE past two decades, Dickie figures, he personally has taught 30,000 people to telemark ski. Instructors trained by NATO have trained still other instructors and skiers, and the videos that Dickie has produced with his cinematographer partner John Fuller (such as *Revenge of the Telemarkers,* a nice mix of goofball entertainment and lucidly explained technique) have spread his voice further. He brings an evangelical zeal to the whole enterprise, a zeal carrying far beyond any mere dedication to athletic prowess, fun-mongering for its own sake, or earning himself a living. He seems to believe that teaching thousands of people to telemark ski will somehow make the world a happier, healthier place.

At the same time, privately, he harbors ambivalence about the telemark category itself. Being chary of neat categories in general, he's chary of that one too. Telemark skiing—how do you define it? With reference to a certain type of turn? Once he would have said so, during his foolish "purist" days, but no longer. A certain type of ski? Definitely not. A certain type of boot-and-binding arrangement that allows the heel to come up and the knee to go down? Naw, that doesn't cover it either. The real essence of Dickie's program is to unify a whole assortment of techniques—virtually every good ski move that has ever been improvised—and use a simplified, versatile set of equipment to transcend all the boundaries separating one stretch of snow from another.

"What we're doing now, you can call it telemark skiing," Dickie says, "but it really is skiing the way it was invented. Put your skis on, go wherever you want, use lifts, don't use lifts. Go to the woodpile, go behind your house, go to the store, across Greenland, go to Denali, go to Aspen." That all-terrain adaptability gives telemark, in his view, a robust future. "I've met a lot of people who used to alpine. I've never met anyone who used to telemark. Because you don't have to give up anything to take it on."

Some devotees prefer other labels for this way of skiing: nordic downhill, norpine, backcountry . . . the list goes on. But Dickie has his own cagey reason for sticking with telemark. "It's really tele*marketing,* in a sense," he admits. "We're using this little fancy turn without heel bindings to get people into a world that's much larger than they even perceive." They come to learn the telemark turn. He and his NATO associates give them that and much more. "But if I called what we do the North American *Skiing Around* Organization, no one would buy our movies or come to our clinics. Right?"

P.T. Barnum himself would understand.

"I WASN'T gonna teach today," Dickie says on Thursday afternoon. But the butter-on-silk conditions have soothed him, and the press of unfinished festival preparations seems forgotten. Having watched me in my own tele gear trying to follow him through a steep mogul field, he finds that he just can't *not* teach.

So I get three personalized pointers from America's most renowned and influential telemark instructor. First, shorten those adjustable poles still farther. Second, keep those shoulders squared down the fall line. His third piece of advice is more cryptic: "Don't be a telemarker." Be a complete skier instead, he means, for whom the telemark turn is only one tool among many. In my case that calls for more parallel turns intermixed with the telemarks, to add quickness and save strain on my legs. He has often given the same unexpected advice to other students, most whom look to him as the guru of telemark in eastern America: "Don't be a telemarker." It's a characteristic sort of tip from Dickie—catchy, arch, and at first impression as mystifying as a koan, but packed full of good sense.

On Friday afternoon, accompanying him again on an inspection of the hill, I see another example of this pedagogic style. A heavyset man in a multicolored coat grinds his alpine boards to an unsteady halt not far from where Dickie has paused. "I don't get forward enough," the man mutters to himself.

Dickie overhears him and hollers a friendly "Nice coat!" by way of hello. Then he offers a suggestion that might help the stranger redistribute his balance. "You need your hands farther forward. Think about keeping your gloves away from your coat." Deadpan, Dickie adds: "Some people say, the farther your gloves are from your coat, the better you ski." Laughing, the man vows he'll try it, as Dickie plunges away down the next pitch.

During my four days of skiing at his side, I witness a number of these Dickiesque bits. "Okay, let's talk about underwear," is how one teaching moment begins. "Let's talk about breakfast food," is another, and with his pole he begins drawing a large fried egg around my feet. Ski with your boots in the yolk—so as not to sprawl out into an exaggerated stride—is the take-away message from that one. For still another, Dickie does his impersonation of the character Lurch from *The Addams Family.* As we weave down a trail, he shouts me a reminder: "Move your pockets back! One after another!" Draw the rear ski backward into my telemark stance, he means, rather than pushing the front ski forward. Each of these cryptic little tips, offered to me or whomever, is calculated to make telemark skiing easier, more economical, and more fun.

"Let's talk about another body part," Dickie says at one point, having watched my fierce efforts to control my hands and my shoulders, to harmonize my eggs and my underwear and my pockets. "The hole under your nose," he says, pointing. "Make sure the corners are turned up."

THE STORM misses, the weather turns sunny, the Sugarbush groomers do their snow-saving work expertly, and the festival hums like a clock. Registration is high, not quite the thousand skiers hoped for but enough hundreds—drawn from New York and Pennsylvania and Massachusetts and Maine and Quebec, as well as the immediate area—to constitute an overwhelming presence. "Safety straps required. Please don't scare the alpine skiers," reads a notice at the bottom of the schedule of events. All over the mountain, throughout Saturday and Sunday, telemark is the prevailing activity and mouth corners are turned up.

A quiet Norwegian named Lars Vik, dressed traditionally in knickers, knee socks, and suspenders, wins the men's slalom with a sleek 34-second run. The clinics—beginner, intermediate, and advanced for adults, as well as beginner and advanced for young kids—take place on schedule, including one intermediate session taught in the comedic mode by Dickie himself. The costuming of some skiers (a Mad Hatter stovepipe here, a propeller-equipped beanie there) helps to remind us all that this is a profoundly frivolous affair, not a ski-industry fashion shoot. The party on Saturday night is a jovial crush of bodies marinated in Catamount beer, loud rock and roll, and the stink of sweaty Capilene, amid which a seemingly tranquil Dickie Hall, separated for now from his clipboard and his walkie-talkie, finds time to dance with his wife. Sunday is another lovely day of corn snow and sun, with the bump contest beginning at 11:00 A.M. on the same steep mogul run where Dickie gave me pointers.

Everything so far has flowed along with just the right blend of careful planning and amiable nonchalance. The only bad management decision of the entire weekend is made by the three bump-contest judges, one of whom is me, when we opt at the last minute to allow each competitor two runs instead of one. Since Dickie has promised us autonomy ("Here are the rules: There are no rules," he explained, "there are no standards, and bribing the judges is okay"), he flinches presciently but doesn't overrule. The disastrous result, however, is that by the time we announce the bump winners, it's too late for Dickie to assemble everyone back at the gentle slope above the lodge for the World Record Attempt Group Telemark Turn. One o'clock has slipped by; in terms of collective energy and focus, also, the moment has passed. The crowd, like the corn snow, is melting away. And so the 20th Annual Festival of the North American Telemark Organization fizzles to a close, with lots of happily exhausted skiers saying their goodbyes, but no big turn. No new record. No linkage of hands, followed by 188 individual pratfalls, as testament to the unifying fellowship and liberating diversity of telemark.

As a responsible party, I feel horrible about this omission.

Never mind, Dickie tells me. Every festival is different. Next year, he says, we'll keep the bumpers to one run. It's okay, David, he says. Not a big deal.

He too is exhausted, emotionally if not physically, and the disappointment shows through.

BUT WHEN I visit him next day at his office in the little loft, Dickie's buoyancy has returned. With a good night's sleep and a quiet morning at home, he has found a more satisfying perspective on the big turn that wasn't. "I think it happened," he says. "It happened at one o'clock. Over two hundred people were doing a telemark turn at that moment." Maybe their separate but simultaneous efforts constitute the *new* big turn, he posits. Telemark skiing has come a long way in a short twenty years. "Maybe we don't *need* to hold each other's hands anymore," Dickie says.

Is this forced cheer? Having gotten to know him much better in four intense days, I don't think so. Besides, so what if it is? We're not talking about brain surgery or global peace, as the man says; we're just talking about skiing, roughly as it was invented and rediscovered. Seated beside the gumball machine, Dickie gives me his unassuming grin. The impresario has faded away, the teacher is on vacation, the clown is resting, and what I see is the philosopher.

# *Eat of This Flesh*

It's an experimental procedure, as far as I'm concerned. The work will take about thirty minutes. Wearing a gray short-sleeved scrub shirt, Dr. Don Thomas performs briskly, with an experienced physician's offhanded precision. His knife is sharp, his preparation has been judicious, and for a while there will be no sign of blood. First he opens himself a beer and passes one to me.

Then he sprinkles a few drops of sesame oil onto the wok and spreads that with his fingers, like liniment rubbed into an achy shoulder. The wok's rusty iron takes a dark sheen. He dices the scallions. He dices the celery, the water chestnuts. The bok choy he dismantles leaf by leaf, chopping each leaf into largish patches. There's no vehement whapping staccato of knife blade against cutting board, since Dr. Thomas isn't one of those operatic chefs, and we talk ramblingly about other things as he works. Meanwhile, mesmerized by his cookery, hungry but ambivalent, I watch every move. He cuts open the red bell pepper, also the green one, and slices them into julienne strips. He pauses for a sip of his beer. He tosses some sesame oil into a skillet on the other burner—a more generous dosage, I notice, than granted to the wok. He cranks up the heat. And now, with a quick move to the fridge, he brings out the meat: two bowls full of mountain lion.

It's prime cut, the stuff that a hunter would call backstrap. To a nonhunter with carnivorous proclivities, it would be recognizable as part of a T-bone. To an anatomically savvy zoologist, paraspinal muscle of *Felis concolor* in the postmortem state. One bowl contains thin-cut medallions. The other, cubes. The lion flesh is pinkish and unappetizing—but not notably less appetizing than raw pork.

The medallions land sizzling in the hot wok. The cubes tumble into the skillet, where Thomas gives them a fast browning and then sets them to a slow, thorough simmer. (Trichinosis, he tells me, is a hazard of undercooked lion. But that doesn't mean that the meat has to be hot-fried until it's as sturdy as Vibram.) The air fills with aroma, and not just the aroma of sesame. The bok choy, the scallions, the peppers, the other items are added to one dish or the other. A splash of soy goes into the skillet, a splash of oyster sauce into the wok. A can of pineapple chunks stands opened and ready. A jar of black bean sauce makes its appearance. While my ambivalence holds at an intermediate level, my hunger ascends toward the ceiling on Cantonese steam.

Already I can tell that Don Thomas, a bright and bristly man with whom I've had a contentious acquaintance, has proven the first of his points: Whatever arguments might be made against the hunting of mountain lions, inedibility isn't one of them.

PERSONAL ethics involves the drawing of lines: I will go as far as this boundary, *here,* but I will not go beyond. I will defend myself against physical menace but only pacifically. I will fight if attacked but I won't kill. I will kill if my family is threatened but I won't aggress. I will squash an earwig in the kitchen but not a beetle in the yard. I will eat plants but not animals. I will eat tuna but not dolphin. I will eat goat but not pig. Fruit but not vegetables. I'm a Jainist, I will harm no living thing—except when I breathe or walk down the street, and then only unintentionally. There's a fuddling welter of such crisscrossing strictures, each observed by its own faction of conscientious people. We all draw our lines in different places, at different angles, and for different reasons, each line's position reflecting a mix of individualized factors that include sensibility, emotion, experience, and taste (in both the broad and the narrow senses of that word), as well as sheer righteous logic. Moral philosophy, unfortunately, is not one of the mathematical sciences.

I will let the butcher do all of my killing. I will destroy habitat but not animals. I will eat stir-fried shrimp, stir-fried beef, even stir-fried elk, but not stir-fried lion. Huh? Not every crisp line represents a triumph of ethical clarity.

Don Thomas draws lines of his own. He will hunt but not with a gun. He uses a bow. He will kill lions and bears, which are thought of by most hunters as trophy-only game, but he refuses to waste the meat. He will focus his hunting efforts toward trophy-size animals, and hang trophy-size heads on his walls, but he declines to enter them in the competitive record book. And even within the narrow domain of bow hunting, he draws a line: traditional weaponry only. Thomas eschews the fancy compound bows—with their pulleys, their lighted sight pins, their release aids, their overdraws, their lightweight arrows, and their other high-tech gimmicks—that have made bow hunting vastly easier and more lethal during the past twenty years. Bow hunting is *supposed* to be much harder than hunting with a rifle, Thomas says. That's a large part of the point. He uses only long bows and recurves, gracefully simple weapons built by craftsmen he knows, from materials such as maple and elm and glue. This is a preference and a scruple, he admits, of which the significance is purely personal.

When hunting, Thomas takes only short-range shots (no more than thirty yards, generally much less) that demand fastidious stalking and raise the likelihood that a hit will be a kill. In autumn he hunts deer, elk, and antelope. Sometimes he goes to Alaska and kills a caribou or a moose. In the depth of winter, with deer season closed and snow piling up in the coulees, he hunts mountain lion. Because lion hunting is impracticable without dogs, he has trained up a good hound. Although he trees a few lions each year and has been at it for almost a decade, he has killed only one. The taxidermed hide of that lion stands in his living room, a handsome but stiff effigy of a splendid animal. Its flesh has long since been eaten. At some point in the future, he says, maybe he'll kill another. In the meantime he will continue enjoying the hunt as a process not dependent on its result. All of these facts are fundamental to Don Thomas's sense of identity. And there's another fact, somewhat more ancillary: He earns his living by practicing medicine in a small town in central Montana.

Thomas is a complicated fellow, not easily captured within categories. As an undergraduate at Berkeley in the late sixties, he was in tune with the zeitgeist, wearing his hair long, protesting the war. Asked if he was a hippie freak in those years, he laughs mildly and answers yes. He missed Woodstock but went to Altamont, tasting the sour dregs of the era. He had majored in English at Berkeley and felt a hankering to write, but then he shifted toward medicine instead. Medicine was a family tradition—his grandfather had been a country doctor in Texas, his father was a distinguished medical researcher who would eventually win a Nobel Prize. After medical school, Don spent two years with the Indian Health Service on a reservation in central Montana, where he felt the reawakening of an old childhood interest: bow hunting. Some of his hunting pals at that time were Indians, but unlike him they felt no inclination toward bow and arrows. He killed his first deer with a bow in 1977, not long after having settled in the town where he lives today.

Also in 1977, he got his first look at the footprints of a mountain lion. "And that was almost like seeing an abominable snowman track," he recalls. The species was so rare—or at least it seemed to be, after decades of persecution—that many ranchers and most town folk lived their lives without ever seeing one.

During the early 1980s Thomas was in Alaska, doing some medicine, some commercial fishing, some bush flying, and a lot of bow hunting. He returned to Montana in 1986 and had another life-changing experience. Hiking in the mountains that summer, as he recounts, "I came over a rise and saw a mountain lion, about fifty yards in front of me." It was his first glimpse of *Felis concolor* in the wild. He was downwind, the lion didn't notice him, and for half an hour he watched the animal working its way along a rimrock. "That was one of the most powerful experiences I've ever had with an animal in the outdoors." This wasn't Yellowstone, or a zoo, this was practically his backyard. "Something clicked," Thomas says. "I knew that I somehow had to interact with these animals."

His interactions took the form of hunting, and though that leap of logic may seem perverse and paradoxical, I can't dismiss it as nonsense, because I remember a similar weird logic in my own feelings about trout. When I first became familiar with wild trout in mountain rivers, they seemed so exquisitely gorgeous, so thrilling, so magically animate, that I wanted to interact—yes, exactly the right word, vague but candid—with them somehow. I wanted to participate in the darting, lambent dynamics of their lives within their environment. Call me doltish and brutal, but I conducted my interactions with a fly rod, stalking the fish, studying their habits, fooling them with devious little baits that I concocted from steel and feather and thread, playing them on a light leader, catching them, and in most cases releasing them, but sometimes killing them and eating their flesh. The trout, of course, never reciprocated my appreciation. They had no desire whatsoever to interact with *me.* But trout are predators too. If they didn't kill and eat insects, small crustaceans, and one another, they wouldn't be susceptible to predation by fly fishermen. That fact is not offered here as an ethical justification for sport fishing. It's just part of the ecological context, worth keeping in mind.

Don Thomas, after his audience with that first cat, became a lion hunter of much the same sort and much the same motivation as me the fisherman. He taught himself some of the requisite skills, and he learned a great deal more at the elbow of a friend called Rosie—full name, John Roseland—who has hunted lions in central Montana for twenty years. Rosie is an amiable fortyish fellow, a seasonal firefighter with long hair and a wrestler's build, who does his own taxidermy and sometimes takes his white toy poodle, as well as his hounds, into the field on a hunt. By Thomas's testimony, Rosie Roseland has forgotten more about mountain lions than most big-mammal biologists will ever know. Thomas and Rosie and just a few other friends now constitute the fellowship of serious lion hunters in their area. They all use traditional archery. Sometimes they practice catch-and-release lion hunting. Sometimes they kill an animal and eat it. If they do kill, the hide is salvaged also and becomes a mounted specimen, though that's not the ultimate object of the process. Rosie has two in his trophy room, including one magnificent tom that went almost 170 pounds, plus another hide now being tanned. The hide in tanning once wrapped the body of a lion that Rosie killed this past December, with Thomas as partner to the chase. They shared the meat. That's the animal whose backstrap is presently featured on Thomas's stove, and Rosie, in recognition of his kill shot as well as for other reasons, has joined us tonight for dinner.

During my own fishing years, I never took a trout to a taxidermist. But I can't offer that as any consolation to the animals I ate. Nor would I argue for any absolute ethical distinction between the killing of a mountain lion and the killing of a trout. If warm blood and fur and mammary glands are enough to set the lion into an exalted category, then we're just back to the old anthropocentric standard of value that has justified such man-against-nature havoc on this planet. In my opinion, the real distinction to be drawn between *Felis concolor* and *Oncorhynchus clarki,* if any, is ecological. Is the native trout common within its ecosystem? If so, then perhaps we can afford to eat it. Is the native lion rare? If so, then perhaps we shouldn't eat it. But even the answers to those conditional questions, as it turns out, are far from simple.

Don Thomas and I have corresponded for a year, in guarded but not uncongenial tones, since he found himself infuriated by something I wrote. It was an essay about the ecological limits faced by large-bodied predators, such as mountain lions, and it contained, in addition to my efforts at scientific explication, some shoot-from-the-hip sarcasm on the subject of lion hunting. To wit: If mountain lions in the northern Rockies are as rare as they *seem* to be, maybe the state fish-and-game laws shouldn't allow hundreds each year to be "shot for the sheer hell of it and converted to rec-room adornments." I also mentioned that, after many years in Montana, I'd never seen one of these animals in the wild. The essay was reprinted by a conservation organization to which Thomas belongs. The organization heard from him, promptly and caustically. One part of the message, as relayed onward to me, was essentially this: If that ignorant yuppie wants to learn something about mountain lions, he ought to look over a lion hunter's shoulder.

Although I'm too old, insufficiently professional, and not quite urban enough to be a yuppie, the shoe otherwise fit, so I thought I'd wear it. I wrote to Dr. Thomas: Yes, thanks, when do we leave? He wrote back, quite affably, and we made plans to go tracking with his dog.

The mountain lion, aka cougar, aka puma and panther, was once the most widely distributed land mammal in the Americas, ranging from the Canadian Yukon to the southern tip of mainland Chile. It has long since been exterminated almost everywhere in the eastern U.S. (though a tiny population holds on in southern Florida, and there are occasional reports of a lion-like apparition in New England). In the West it has fared better, not for lack of enemies but most likely because the conditions of habitat and the sparseness of human settlement permitted its survival.

From the pioneer days until just thirty years ago, mountain lions in the western states were slaughtered remorselessly, to the point of eradication in some parts of their range and severe reduction in most others. Stockmen hated them, and bounty hunters were paid to shoot as many as possible. During the first three-quarters of this century, the *documented* tally alone came to 66,665 dead lions. But several factors seem to have saved the western populations from being killed off entirely. Mountain lions are elusive enough, under most circumstances, to avoid contact with humans. They have a relatively high rate of reproduction. Their food supply—mainly mule deer and whitetails—has cycled upward to the point of abundance. The payment of bounties was finally ended. The indiscriminate poisoning of predators with Compound 1080 was curtailed. In 1965, Colorado became the first state to reclassify the lion as a game animal. Other states followed, including Montana in 1971. One effect of this last factor, the change from varmint status to game-animal status, was that it gave hunters a stake in the conservation of *Felis concolor*.

In the years since I wrote the essay that angered Don Thomas, prevailing wisdom on the population status of mountain lions in the northern Rockies has changed. Possibly their actual status has changed within those few years too, though the current trend goes back several decades. And my own awareness of the particulars, which was shallow before (partly because my argument was focused elsewhere, partly for less valid reasons), has deepened at least slightly. This is not a retraction, then, but it is an admission of incompleteness. Most biologists now seem to agree that populations of *Felis concolor* in the northern Rockies have increased dramatically within the last thirty years. A pleasant surprise, running counter to most other trends in the struggle to conserve jeopardized species: We've got more lions than we used to, it seems. The kill rate attributable to sport hunters is probably no direct threat to the demographic health of the species.

Other things are. Some stockmen continue to argue for government-funded eradication of mountain lions. And home-in-the-country fever continues to afflict more and more people, especially the recent arrivals from faraway cities, who imagine that a cabin on forty acres in the foothills, a Jeep Cherokee, a pair of hand-tooled boots, and a dip of Skoal will give them some sort of spiritual rebirth. This home-in-the-country fever is probably destroying as much wild landscape, as much lion habitat, as any other factor. It's more permanent than clearcutting, it's more heedless than Compound 1080, and it's very damn hard to control. It's turning Montana, for one instance, into a cross between Walden Pond and Levittown.

Accidental encounters between humans and lions are also increasing, and a few of those encounters have yielded human fatalities. This sort of conflict will bring the lion new trouble. As *Felis concolor* becomes more numerous, more constricted by human encroachment on its habitat, and possibly (for reasons still unclear) less shy, it will risk triggering a campaign of retributive control. Don Thomas has noticed intimations of that in his own community. "People are just talking about it now. But the first time a lion does something here—whacks a kid, or whatever it happens to be—there's gonna be a tremendous amount of pressure brought to bear to reduce the mountain lion population." He hopes it doesn't happen. He takes joy in knowing that the woods are full of lions.

One morning last winter he tracked a female and her three yearlings in a coulee within walking distance of his house. After treeing them with his dog, he took a few photos and then let them be. It wasn't the day for a kill. When word leaked into town that Dr. Thomas had had four lions at bay—virtually in the suburbs, for Christ's sake—and had intentionally allowed them to escape, some of the good citizens were unhappy.

Some were merely puzzled. Is this guy Thomas a lion killer, or isn't he? This answer to that one is yes.

In the depth of winter, with deer season closed and snow piling up in the coulees, Don Thomas and I spend two days together, looking for mountain lions and common ground. We leave town each morning before dawn and drive far into the mountains with his Bluetick hound, Drive, in the back of the pickup. Each day we cover a long stretch of deserted roads, fighting through drifts, scanning the borrow pit and the shoulder for fresh tracks. We hike and we ski through some lovely country. We investigate a handful of his favorite spots—places where, as he knows from experience, a lion is likely to cross—while Drive, his nose high and his nostrils flared, reads the aromatic braille of the breezes. The wind is up, with another blizzard reportedly on its way. The air is just cold enough to focus a person's brain. Lunch freezes in my pack, and so does the canteen. As we ski, hike, and drive, we also talk.

Don mentions certain cases of "bad logging" that are ravaging habitat in these mountains, and of "bad mining" that are leaving heavy metals in some of the streams. At another moment, he adds that "one of the biggest impending tragedies" in the struggle waged by conservation organizations, to his mind, "is the polarization between the hunting and non-hunting factions of those groups." This polarization divides resources, it divides people, it wastes time and money, he says. Until hunting and non-hunting conservationists can find the pragmatic wisdom to accommodate each other within the larger fold, "the developers and the miners and the loggers have gotta be laughing all the way to the bank." We also discuss home-in-the-country fever, about which he shares my concern. To both of us it seems maddeningly obvious that, if everyone who purports to cherish wild landscape decides that he or she must own and live on a chunk of it, there won't *be* any more wild landscape. But how do you tackle that problem, when the new home-in-the-country developments are full of well-meaning people—conscientious hunters, ethical vegetarians, Sierra Club members—some of whom you personally know and love? Don has no answer, nor do I. In two days of searching, we don't see a single lion, either. But he has convinced me that they are here to be found. Maybe the hard answers are too.

With no lions showing themselves, my interaction with *Felis concolor* is limited to the culinary mode, God help me. Back at Don's house, I stand beside his kitchen island, in my hungry ambivalence, while he stirs those medallions in the wok. I watch him dump the pineapple chunks and the julienned peppers into the skillet along with the cubes of flesh and then bring that dish abubble in a pungent gravy. I listen to Rosie describing his own day's effort—he cut the tracks of four different lions, followed them for a few hours in befuddling circles on a ridge, never spotted an animal in the flesh—and his own paradoxical feelings toward the species. Some of his neighbors have suggested to Rosie, as they have to Don, that if he's such a lion hunter, he ought to exterminate the local population. "Kill them all?" says Rosie. "Hell, that's the last thing I want to do." It's been a good day for Rosie today, because he went into the woods and worked up a sweat and found evidence that he was in the company of *Felis concolor.* Let's eat, says Don.

He sets two steaming platters before us on the table: sweet-and-sour mountain lion, and mountain lion in black bean sauce. He pours a nice chardonnay. Before anyone has touched a serving spoon, he raises his glass. "To the animals," says Don Thomas. This is tradition between him and Rosie, who raises his own glass. "To the animals." I raise mine. To the animals. And then we eat. It's the sacrament of accommodation and alliance.

[IV]

# *The* HEART

# *The Swallow That Hibernates Underwater*

We all make our deals with life. We do it invisibly, sometimes unconsciously, and alone, without benefit of collective bargaining. We come to terms.

And the terms are in every case different. Some of us hold out for more, for better, when others would settle. Some of us settle when others would hold out. We leave home, or we marry early, or we enlist, or enroll, or audition, or hunker down into a job; then we jump through the hoops, or fly off the tracks, or sell out if there happen to be buyers; we invest years, gamble dollars, marry late, light out for the territories, raise a fine family, raise hell, declare Chapter Eleven, buy a Harley-Davidson on whim and roar off down the highway clad in leather, save, scrimp, drive a fifteen-year-old Nova with rust, sacrifice joyously, piss it away, loiter, maunder, hold the course, see our child graduate, keep faith with our commitments, sow oats in a high wind, marry often, travel great distances in search of a place to call home, lose big, win big, harbor regrets, fulfill finally our one wildest dream; or, alternatively, we don't. The compoundment of it all—what we've done and endured, what we've left undone and refused to endure—is our individualized deal, also called sometimes by the more august and despairing word *fate.* There is no fate. There's a lifelong balancing-off between the possible and the all-too-likely, resulting in a succession of half-chosen arrangements, of which the last is burial or cremation. We can even specify that our ashes go into a silver urn, or into a mountain river. Then our deal-making is over, unless we've devised a nefariously prescriptive will. Your deal is unique to you, mine to me, but we share the process. During the early and middle decades of his adulthood, more than two centuries ago, an unassuming English clergyman named Gilbert White was arriving at a deal of his own.

This man, like you and me, had to reconcile the tension between what he might want out of life and what, on the other hand, he was willing to accept.

Gilbert White is famed as the author of *The Natural History of Selborne,* one of the most persistently cherished books in English literature. According to a recent tally, White's *Selborne* is the fourth most published title (as figured by number of different editions) in the language. White himself has been celebrated as the grandfather of ecology and as the paradigm of the natural history essayist. Despite the ponderousness of his reputation, he was in fact an exceptionally keen observer and a nifty writer. He made a great difference, at the dawn of modern science, by studying the lives and habits of animals, instead of merely their dried carcasses. He published only the one little volume, and did almost nothing else even faintly impressive, but his book is full of small insights and charm and secret significance, as potent in its own way as *Walden.*

For right now, though, let's set the book and its impact to one side. Maybe you already know *Selborne.* If not, you can read it someday in a busy airport, when you need a tranquil counterpoint to reality, and make your own judgment then. Two other facts about Gilbert White are more intriguing at the personal level—to me, anyway—than his place in literature and in science.

Fact one: He lived most of his life, and died, in the same little village where he'd been born.

Fact two: Despite fifty years of close study, he never abandoned his belief in the hibernation of swallows.

* * *

THE HIBERNATION of swallows is a misapprehension as old as Aristotle, who offered it in his *Historia Animalium* in the fourth century B.C. A great number of birds "go into hiding" rather than migrating in winter to warmer locales, according to Aristotelian pronouncement. "Swallows, for instance, have been often found in holes, quite denuded of their feathers," he wrote, adding that ouzels and storks and turtle doves also went torpid and hid themselves. An ouzel, that peculiar semiaquatic bird actually capable of walking on the bottom of a river, might even be imagined to hibernate underwater. A stork would presumably need a hollow tree.

There were other nuggets of erroneous biological dogma in Aristotle's work—for instance, the bit about eels being born from earthworms and earthworms being spontaneously generated in mud—but the hibernation of swallows was a fancy that endured. Two millennia later, in Gilbert White's era, the great systematizing biologist Carolus Linnaeus may have believed it himself. One of Linnaeus's more conspicuous students, A.M. Berger, definitely believed it, mentioning the notion in his *Calendarium Florae* as though it were certified fact. Berger's version of swallow hibernation was the radically wrongheaded subaqueous one. Each season had its reliable signals, and early September, Berger observed, was when swallows went to hibernate underwater.

In *The Natural History of Selborne* White cited Berger's claim, admitting that he was tempted to believe it himself. Underwater hibernation seemed to jibe with what White had noticed on his own: that swallows and their near relatives, in mid-autumn, with the young fledged and the nests abandoned, tended to gather in nervous flocks around ponds and rivers. "Did these small weak birds, some of which were nestlings twelve days ago, shift their quarters at this late season of the year to the other side of the northern tropic?" he asked. "Or rather, is it not more probable that the next church, ruin, chalk-cliff, steep covert, or perhaps sandbank, lake or pool (as a more northern naturalist would say), may become their *hybernaculum,* and afford them a ready and obvious retreat?" This wasn't so much a rhetorical question as a genuine uncertainty within his own mind: Was it more probable, or not? He couldn't decide. *The Natural History of Selborne* took him eighteen years to compose (partly because he insisted on padding it out with a pedantic section on historical antiquities, which has been mercifully omitted from some editions), based on a long lifetime of watching and rumination. But he never did settle the question of wintering swallows.

His eyesight and his knowledge of birds pushed him toward one answer, I think, while his heart preferred another.

SELBORNE is a tiny, ancient village set among the hills and meadows of Hampshire, about forty miles southwest of London. In White's time, those forty miles represented a long journey by coach from one world to another. The lanes of Selborne are cut deep as canals by centuries of traffic and erosion. In the churchyard is a yew tree, huge in girth and guessed to have stood for perhaps more than a dozen centuries. Just beside the church is the vicarage, where Gilbert was born in 1720. His grandfather was vicar of Selborne.

In basic outline, Gilbert White's life appears simple and happy and sweet. From the age of about nine, after his grandfather's death, he was raised in a sizable house called The Wakes, just across the village green from the vicarage. He was an outdoorsy boy who planted trees and occasionally made notes on his natural history observations. He went off to boarding school, then to Oxford, and in his mid-twenties took deacon's orders, which made him essentially a licensed clergyman in search of a job. During his thirties he traveled widely around England, visiting friends and extended family, seeing the countryside, and serving in some temporary posts as a fill-in churchman. Always, even during those years, Selborne remained his true home and retreat. In 1760 he returned there permanently. He accepted a modest clerical position nearby. He lived the rest of his life at the old family home, The Wakes. He became a serious gardener and began keeping a horticultural diary, quite terse and businesslike at first, which gradually evolved into the journal of a full-hearted naturalist. He never married or, apparently, even came close. According to a later biographer, Gilbert "had but one mistress—Selborne." Another scholar of White's life and work, Richard Mabey, has written: "He never scrimped his clerical duties, but with only a few dozen marriages and burials to attend to a year, had plenty of free time to pursue his natural history." He watched birds. He recorded the seasonal timing of their activities in his journal, year after year. He paid attention to crickets and slugs and hedgehogs and the weather and the hibernating rhythms of an old pet tortoise named Timothy. He raised cantaloupes, succulent enough to make a preacher proud. He studied the barn swallow, *Hirundo rustica*, and certain similar species (martins and swifts) with particular devotion. Eventually he wrote his book, styling it as a series of information-filled letters to two other men, both of them naturalists better traveled and better known than he. The first edition was a modest success. Forty years after his death, the book became surprisingly, voguishly popular. Historians now talk about "the cult of Gilbert White and Selborne." Readers have belatedly recognized that this unprepossessing man (of whom there seems to be no surviving portrait) found great scope and great wisdom, as well as beauty and peace, within the boundaries of his little village. He lived to age seventy-two. Needless to say, this outline leaves a hell of a lot out.

It leaves out, among other things, the tension between what he may have wanted from life and what he got.

One sample of that tension, from the larger pattern: Gilbert, unlike his grandfather, never became vicar of Selborne. He couldn't. For reasons that must have seemed adequate to a nineteen-year-old, he had gone to the wrong college at Oxford—Gilbert was an Oriel man, whereas the vicar's position at Selborne was restricted, by certain archaic regulations, to graduates of Magdalen. Disqualified to be vicar, what he eventually became instead was curate of Selborne, and the difference is more than semantic. Within the Anglican hierarchy of White's time, a curate was just a delegated assistant who functioned as acting pastor. The curacies were often short-term assignments, meagerly paid, like a non-tenure-track lectureship at a modern university. A vicar was a salaried professional; a curate, subcontracted under a vicar, was a liturgical flunky. And even the Selborne curacy came to him late in life. Throughout most of his middle age, though he lived in Selborne, he held a curate's position not for Selborne itself but for one or another small village in the vicinity, to which he commuted by horse. Old friends goaded him for decades: Gil, you shirking doofus, with a little push you could land a respectable Oriel vicariate somewhere else. He declined to push. At one time, when he was younger and more energized, he had tried for a good position through Oriel and been rebuffed. He wouldn't try again. He had money enough to get by. Selborne was familiar and safe, and it was home. He loved the meadows and woods. If he had any broader ambitions, they were secondary to his sense of place.

Another sample: In 1763, when Gilbert was a 42-year-old bachelor whose juices still flowed, three young sisters swept into Selborne for a summer visit and caused him some delicious perturbation. Their names were Anne, Philadelphia, and Catharine Battie. They were in their late teens and early twenties, and (as nicely said in the wonderful White biography by Richard Mabey, to which I'm indebted for most of these facts of personal history) they were "rich, flighty and attractive. They fizzed about the village for two months, and left a perceptible dent" in Gilbert's composure. There were balls and picnics and other sorts of gently flirtatious shenanigans. Catharine was the sister who caught Gilbert's eye. Her own attention seems to have flickered more on Harry, Gilbert's much younger brother. Gilbert himself, after all, was a middle-aged man of no particular forcefulness, a bachelor past his prime, and not even a vicar but a curate. The summer interlude was fun but not serious—at least not, evidently, to the Battie sisters. They left Selborne in August. On November 1, Gilbert wrote a poem, gloomy with autumnal images and dedicated "To the Miss Batties," which ended:

> *Return, blithe maidens; with you bring along*
> *Free, native humour, all the charms of song,*
> *The feeling heart, and unaffected ease,*
> *Each nameless grace, and ev'ry power to please.*

But they never did.

Still another sample: In 1768, White sent (indirectly, through a mutual friend) an invitation to Joseph Banks, the celebrated and wealthy young naturalist who was about to leave on a round-the-world voyage of exploration with Captain Cook. Would Mr. Banks and the mutual friend care to visit Mr. White in Selborne? If Banks will just do him the honor, White promised, "he will find how many curious plants I am acquainted with in my own Country." It must have sounded faintly pathetic. Getting no acceptance, White wrote directly to Banks, a polite but mopey letter in which he complained that, if Mr. Banks and other busy colleagues wouldn't visit him, "I must plod on by myself, with few books and no soul to communicate my doubts or discoveries to." Joseph Banks sailed away on the *Endeavour.* He would visit Tierra del Fuego, and Tahiti, and New Zealand, and Australia (where he'd discover, among other things, kangaroos), but not Selborne.

Gilbert White was a stay-at-home guy during an age when the great naturalists made great expeditions; and he knew it. Broad travel, the collection of exotic observations and specimens, seemed fundamental to the enterprise. Linnaeus had gone to Lapland. Banks, even before the Cook voyage, had done fieldwork in Newfoundland and Labrador. Pehr Osbeck sailed to China. Johann Gmelin got to Siberia, and Carl Peter Thunberg to imperial Japan, eventually publishing his *Flora Japonica.* Darwin and Huxley and Hooker would later make crucial voyages of discovery. Henry Bates would bring important data and insights back from the Amazon. Alfred Russel Wallace would wander the Malay Archipelago for eight years.

Gilbert White, as he grew older, as he settled more rigidly the terms of his life, traveled less and less. Even the familiar roads of southern England got to be too much for him. By his own account, he was prone to horrible coach sickness. But we shouldn't assume that he achieved obliviousness to what he was missing.

I don't claim that these particular deprivations—of a vicar's position, of Catharine Battie's affection, of the inspiriting companionship of Joseph Banks and other naturalist colleagues, of the chance to go off on a great expedition himself—were the four big facts of his life. What I suspect is that they were representative.

SWALLOWS, like martins and swifts, feed on insects taken in flight. Their amazing agility on the wing is a prerequisite to this dietary strategy. They cruise, they dive, they swoop, they swim through the air, gathering small mouthfuls of gnat and mosquito and beetle. Aquatic insects such as mayflies and caddisflies—which emerge from the water's surface as winged adults, often in synchronous events during which one species or another fills the air with its blizzard-thick multitudes—are especially convenient food for these birds. So they tend to congregate around rivers and ponds. They are also drawn to villages and small towns, where human-made structures with overhung roofs and rafters offer good sites for their nests. But if the insect density in a certain locale isn't high, or if the insects aren't taking wing at some particular time of year, then swallows and martins and swifts can't afford the metabolic cost of their habitual swooping and diving. They can't make a living there. So they migrate.

Besides being graceful, they have stamina. They travel long distances out of the north in order to winter in warm, buggy places. Swallows from Siberia go to Sumatra. Swallows from Canada go to South America. Swallows from Europe cross the Mediterranean and the Sahara. Although the movements of discrete populations aren't easy to track precisely, modern bird-banding work suggests that the British population of *Hirundo rustica,* the barn swallow, probably spends Christmas in South Africa.

Gilbert White never banded the birds of Selborne. Sometimes he shot them and dissected them. He peeked into their nests. Mostly he watched them, acutely but lovingly, from a respectful distance.

In *The Natural History of Selborne,* he wrote: "The *hirundines* are a most inoffensive, harmless, entertaining, social, and useful tribe of birds: they touch no fruit in our gardens; delight, all except one species, in attaching themselves to our houses; amuse us with their migrations, songs, and marvellous agility; and clear our outlets from the annoyance of gnats and other troublesome insects."

Then he *did* believe in swallow migration—that notably amusing attribute—as opposed to swallow hibernation? Yes and no. Elsewhere in the book, White voiced his equivocating opinion "that, though most of the swallow kind may migrate, yet that some do stay behind and hide with us during the winter." Joseph Banks had gone off to remote places that White would never see; the Misses Battie had gone off; so many others had gone off and left him behind—including, each autumn, most if not all of the British swallows. But maybe some of those birds lingered secretly in Selborne, mitigating the wintry isolation of a poor bachelor curate.

He imagined that they might "lay themselves up like insects and bats, in a torpid state, to slumber away the more uncomfortable months till the return of the sun and fine weather awakens them." Where exactly did they hide? Not in their old nests; he had checked. But he recalled the curious affinity they seemed to have for rivers and ponds, which caused him "greatly to suspect that house-swallows have some strong attachment to water, independent of the matter of food; and though they may not retire into that element, yet they may conceal themselves in the banks of pools and rivers during the uncomfortable months of winter."

Where were the swallows of December? He was honest enough to claim no final answer.

His book was published in 1789. He died in 1793, at his house in the village. His death came in June of the year, so we can assume that the swallows were back from Africa, nesting outside his window.

Gilbert White's reputation is based partly on the fact that he was an extraordinarily fine observer, a painstaking empiricist who relied on his own eyes and ears, not on secondhand anecdotes and theoretical preconceptions. His mistake about swallow hibernation was an uncharacteristic lapse. He had ulterior reasons, I think, for keeping the idea alive. These reasons pertain not to the ecology and behavior of *Hirundo rustica,* of course, but to the natural history of the human soul. The swallow that hibernates underwater is a creature called yearning.

# *Trinket from Aru*

On a back street in the town of Dobo, principal port of the Aru Islands, one hundred miles offshore from southwestern New Guinea, a man offered to show me a packet of pearls. Although I wasn't looking for pearls, I hadn't yet found what I *was* looking for, so why not? Along the sea-trading routes that stretch westward through the Moluccas, beyond Sulawesi, even as far as the great cities of Java, encompassing that region once known as the Malay Archipelago and now comprising Indonesia, little Aru has been anciently famed for its pearls.

The man poured his half-dozen small globes onto a dark table and invited me to inspect. I rolled one between my fingertips, then another, peering astutely, as though I knew pearls from Shinola. They were no bigger than chokecherries, imperfectly round, but their lumpy pinkish-gray surfaces did catch the light prettily, and they had the aura of unmistakable authenticity. Also, they were wholesale. They had never been handled by a jeweler. This particular back street was labeled Jalan Mutiara ("Street of Pearls"), so clearly I was dealing near the source, maybe just one intermediary away from the Aruese pearl-diver who harvested these little treasures, maybe just two from the Aruese oysters that made them.

*Berapa harga?* I asked the pearl-trader. How much? He quoted me a price in rupiahs—*tiga puluh ribu,* thirty thousand, roughly equal to fifteen dollars—and though undoubtedly he had inflated that price at first glimpse of the white man in the goofy sun hat, I'm sure they were easily worth it, as mementos if not as sheer merchandise. They would be nice for my wife, yes? Well, if she happened to be the pearl-loving sort, I suppose they'd be nice for her, yes. The man waited for me to begin bargaining.

Thanks but no thanks, I said, and walked on. I was too distracted with my own primary mission to commit myself, just then, on a souvenir.

My own primary mission was to charter a boat and hire a guide for a trip to the backcountry of the islands. I had come hoping to see *Paradisaea apoda,* the greater bird of paradise, a creature for which Aru has been anciently famed in regions far beyond Sulawesi and Java.

DOBO is an old-fashioned trade entrepôt, not a nexus for ecotourism. More than a century ago Alfred Russel Wallace, who traveled for years through the region making natural-history observations and collections, wrote: "Rattans from Borneo, sandalwood and bees'-wax from Flores and Timor, tripang from the Gulf of Carpentaria, cajuputi-oil from Bouru, wild nutmegs and mussoi-bark from New Guinea, are all to be found in the stores of the Chinese and Bugis merchants of Macassar, along with the rice and coffee which are the chief products of the surrounding country. More important than all these however is the trade to Aru, a group of islands situated on the south-west coast of New Guinea, and of which almost the whole produce comes to Macassar in native vessels." Macassar was the chief city of Celebes (as Sulawesi was then called), and the Bugis were its seafaring traders, for whom a voyage out to Aru under power of the westerly monsoon winds could be the central event of a lifetime, like a Moslem's long pilgrimage to Mecca. "The trade to these islands has existed from very early times," Wallace added, "and it is from them that Birds of Paradise, of the two kinds known to Linnaeus, were first brought."

The two kinds in question were *Cicinnurus regius,* a small scarlet species commonly known as the king bird of paradise, and the greater bird of paradise, *Paradisaea apoda.* Linnaeus himself had described and catalogued them in 1760. But what Linnaeus had worked from were dried skins, and neither he nor the other European naturalists of that era had ever seen one of these birds alive. The specimens that they did see were few, expensive to acquire, bogglingly decorative, anatomically incomplete, poorly understood, and all the more fascinating in consequence. The same pair of species enticed Wallace himself to make the journey out to Aru in 1857, traveling the monsoon route by prau from Macassar, and a half-crazy compulsion of my own had provoked me to follow him. I wanted to see for myself whether *Paradisaea apoda* had survived in the wild.

But there was no prospect of getting what I was after—a boat, a guide, a glimpse of a wild bird—without immersing myself in the moil of commerce. The town of Dobo, like Singapore or Chicago, is a mercantile construct. By that I mean: It's no more than a convenient spot at which sellers have traditionally met buyers, not a site chosen for permanent habitation on the merits of its spring water or its copper resources or its soil. The town covers a low arm of beige coral sand on one of the smallest and biologically least interesting islands of the Aru cluster. It serves simply as a marketplace for the precious commodities from the other islands. Exchange is its fundamental enterprise, by which values are set and human relations are driven. To find my way in this town and my way out of it, therefore, I went through the motions of shopping.

In a storefront near the dockside warehouses, I was offered a cassowary egg, blown empty and mounted on a stand. At *sepuluh ribu,* five dollars, it represented a good bargain for the scruple-free tourist, though probably a bad one for the local species of cassowary. Dobo was full of such bargains: living or once-living rarities to be had at factory-outlet prices. I was offered a pearl-shell ashtray. I was offered the shell of a chambered nautilus. I was offered the delectation of dried sea cucumber in three different sizes: breakfast sausage, bratwurst, and Polish. With a little searching and dealing, I'm sure, I could have carried off a supply of premium shark fin and enough cave-swiftlet nests for a tureen of bird's-nest soup. Down a back alley in the tidewater sector of town, where the houses are thatched shacks on stilts and the yards are sumpy with garbage, I was furtively offered two gorgeous green parrots. The parrots were miserable but alive, shackled with twine to a perch, yelping each time they fell and dangled. I asked the parrot man the same thing I'd asked the pearl trader: *berapa harga?* He smiled in the confidence that he had something I wanted. *Tiga puluh lima ribu,* he said, adding pointedly, *satu.* Thirty-five thousand, for one. Where exchange is the fundamental enterprise, life itself has an asking price, and for a parrot, it's seventeen dollars. The man's face opened in expectation of haggling.

Thanks but no thanks. I was morbidly curious but I wasn't buying.

And then a sweet-faced, half-Papuan fellow I met in one of the warehouses asked me: Would I be interested to see his *cenderawasih*?

I would be fervently interested. I knew barely enough of the language that my ears twitched at *cenderawasih.* It's the Indonesian name for the greater bird of paradise, *P. apoda.*

THIS FELLOW led me down lanes, across catwalks, to a rough shack on stilts above the garbage-strewn tidal muck. Against the back wall of an unlighted room stood a homemade cage. Draped over it lay a mat. On the adjacent wall hung a garish color print of the crucified Christ. There was no furniture. The room quickly filled with children, women holding babies, neighbor men, all of whom evidently wanted to ogle the pale outlander who wanted to ogle the bird. I can't say much for what they saw but, when the mat came away, I saw a spectacular animal.

It was a male *P. apoda* in full breeding plumage—yellow head, auburn body and wings, black muff around its bill, a patch of green iridescence at its throat, and, cascading backward from under each wing, the long willowy golden plumes that in this species serve as the banner of sexual display. It was the size of a crow, but a crow dressed as Egyptian royalty. The floor of the cage was littered with birdshit and half-nibbled bits of banana. For a water dish, there was the bottom half of a Sprite can, jagged-edged and dangling from a wire. The bird jumped around nervously as I stooped close to gape.

After some minutes, I asked the inevitable. No crude *berapa harga?* this time; I tried to be discreet. What might be the value of such an animal? I expected an extravagant figure because the species is rare as well as resplendent, and because the trade in birds of paradise is strictly illegal. A robust *cenderawasih,* at Dobo, I learned, might bring 250,000 rupiahs.

For about the same amount, hours later, I had my boat and my guide. We left town on the afternoon tide.

OF ALL the forty-two bird-of-paradise species presently known in the world, *P. apoda* is in some ways archetypal. Like many of the others, it shows extreme sexual dimorphism, the males gaudy, the females drab. With its green throat, its red breast, its golden pectoral plumes, a male *P. apoda* embodies the result of what Darwin himself labeled *sexual selection:* seemingly whimsical evolutionary modifications driven not by the practicalities of survival but by competition within one gender for access to mates. Also like many other birds of paradise, *P. apoda* practices lekking behavior, a form of courtship in which males congregate in a single area to show off their sexual worthiness before an audience of discriminating females. The genus to which it belongs, *Paradisaea,* includes five other auburn-breasted species, all of which closely resemble *apoda* (plus one species, the blue bird of paradise, that looks much different). The red bird of paradise, the goldie's bird of paradise, and the raggiana bird of paradise are distinguished from the greater bird of paradise by red pectoral plumes instead of yellow ones; the emperor bird of paradise has white pectorals and a larger patch of iridescent green; and the lesser bird of paradise is a very close match to the greater. In the midst of this crowded genus, the greater bird of paradise is the one that still bears the telltale Latin name, *apoda,* meaning footless, as a vestigial reminder of the early misapprehension (by Linnaeus, who named it, and most of the other Europeans who saw only dried, boneless skins) of the paradise birds as unearthly beings that spent their whole lives in the air without ever descending to perch. Since the specimens reaching Europe were footless, the living birds were imagined as footless too.

Humans have probably treasured these creatures for their plumage—hunted them, skinned them, invented magical names and legends to do justice to their improbable splendor, bought them and sold them—almost as long as humans have known them. The traditional Huli people of the central New Guinea highlands still use the display feathers of several different species to adorn their own ceremonial costumes. The early Malay traders spoke about *manuk dewata*, according to Wallace, meaning "God's birds." Their vogue as an item of long-distant export, from the New Guinea region (including Aru) to the wider world, is a more recent phenomenon; but when it caught on, it caught strongly. The first skins seen in Europe were brought back to Spain by surviving crew members from Magellan's voyage in 1522. Portuguese travelers also encountered a few, and knew them only as *passaros de sol*, or "birds of the sun." A Dutchman named Jan Huyghen Linschoten used the "paradise bird" label in 1598, while floating the story that they were footless, ethereal beasts permanently aloft in the heavens. In 1824, a French explorer became the first European to view them in the wild, and then Alfred Wallace went to great risk and trouble for his own chance in 1857.

Wallace himself was a commercial collector and trader as well as a naturalist. He financed his travels throughout the Malay Archipelago by shipping specimens back to an agent in London, for sale both to museums and to dilettantish fanciers who kept their own private natural-history collections. He shipped more than a few birds of paradise to that market. Wallace wasn't greedy, his purpose was noble, he made huge scientific contributions and wasted nothing; but if he committed the same depredations today, the World Wildlife Fund would put his face on a wanted poster. His collections from Aru in particular were the richest and most lucrative of his whole eight-year expedition, netting him more than £1,000. He took rare pigeons and flycatchers and parrots, at least one black cockatoo, a racquet-tailed kingfisher, a cuscus, a huge green-and-black birdwing butterfly that especially thrilled him, and a large number of other butterflies, beetles, and small birds, as well as several good skins of *Cicinnurus regius*, the king bird of paradise, and about two dozen *P. apoda.*

Then, during the late Victorian era, bird-of-paradise feathers became a fashion craze in Europe for the adornment of ladies' hats. Ironically, although Wallace expressed concern that the two Aru species might be jeopardized, he himself may have been partly responsible. He had opened the trade channel to London, and his writings had publicized the birds' extraordinary beauty and preciousness, which probably served to prime the demand. In the following decades, the toll became ecologically significant. At the end of the nineteenth century, by one account, three thousand bird-of-paradise skins were being exported each year from Dobo. A few bird lovers of the day were perspicacious enough to realize that, if the harvest continued at any such rate, the wild populations were doomed. And the prospect of their doom led to another episode of extraction—this one performed, paradoxically, with the best of intentions.

In 1909, four dozen individuals of *P. apoda*, captured alive from the forests of Aru and carefully transported across the world, were released on a tiny island in the West Indies. The hope was that they would thrive in this alien but protected habitat and establish a breeding population in exile, safe from the plume-hunters at home. The new island was Little Tobago, a three-hundred-acre nub of neotropical forest just off the coast of Venezuela.

The translocated population did survive—or at least, linger—but it certainly didn't thrive. By 1966, only seven birds could be found. Of those, four were males in full plumage and the other three were either adolescents or females. Their gene pool was tiny and their social system was fractured. They showed no evidence, that year, of having produced a single successful nest. But the four males on Little Tobago continued performing their courtship displays, like the empty rituals of a Cargo Cult or the hymns of a dead religion.

The Little Tobago population represents a peculiar footnote to the long history of how Aru has been stripped of its precious commodities. For centuries divers had taken its pearls and its pearl shells and its sea cucumbers; fishermen had taken its shark fins and its turtle shells; nest gatherers had taken its edible swiftlet nests; hunters had taken its birds of paradise; and the traders who converged at Dobo had taken them all. Now an unexceptional island on the far side of the world, Little Tobago, took a large share of its thunder. Beginning soon after the translocation of birds and continuing throughout most of this century, ornithologists went there, not to Aru itself (where the species had survived the plume-hunting craze but which was still so impossibly remote) for their field studies of the greater bird of paradise. Among the more recent of those ornithologists was a graduate student named James J. Dinsmore, who gathered nine months' worth of observations on Little Tobago and then published a scientific paper.

A copy of Dinsmore's study was stashed in my baggage when I landed at Dobo. One sentence was underlined, because it struck me as mildly shocking. After listing the earlier field studies done on Little Tobago, Dinsmore had noted passingly: "No one since Wallace has described the display of *apoda* from the Aru Islands."

So HERE's my description:

Deep in the forested interior of Aru, half a day's journey by pearl-diving boat from the commercial hubbub of Dobo, then some miles farther by dugout canoe and foot trail from the nearest small village, on the crest of a ridge, stands a towering tree. Its broad, open crown includes some stout horizontal limbs, bare and straight, that are just right for lekking behavior by *P. apoda.* With an Aruese guide, at whose village I had slept and in whose canoe I had ridden, I reached the site just after dawn. We could hear the birds before even seeing them. They sounded like a barnyard full of rioting ducks.

Four or five males, in full plumage, were flaunting their sexual merits before an audience of interested females. They were eighty feet up, dancing their brains out along the lekking limbs, but by sprawling flat on my back at the base of the tree with binoculars locked to my face, I got a good view. The males skrawked in chorus, they flapped, they charged back and forth, each one making maximal racket and showing maximal flash. Their competitive display seemed to resolve itself into brief one-on-one skirmishes. A male ran south on this limb, flying his colors, while another male ran north on that parallel limb nearby. Their calls rose in pitch and in pace toward a mutual crescendo: "Wok, wok, wok, wok-wak-wak-wak-wak-weekweekweekweek!" Then, after passing like jousters a few times, they suddenly froze in position—wings forward, heads down, the display plumes from beneath each wing cranked upward like billowing fountains of chardonnay—and went silent. They held that position for long seconds while the females studied them. It was a stunning interlude in a raucous performance. The males strained; the females gawked; the moment was charged with sexual frenzy and Darwinian import. From my viewpoint, the motionless males were backlit to shades of luminous auburn and yellow against patches of bright morning sky. The air of the forest itself seemed aflood with avian hormones.

Lying there beneath the great tree, I watched this pageant for ninety minutes, pausing only to unfog my glasses and scribble notes, while the Aruese chiggers burrowed holes in my skin.

CHIGGER bites were the least of the costs. I was trading a large pile of money and a month of tedious travel, mostly aboard a Dutch-chartered cruise ship, for this brief session of extraordinary bird-watching. I'd have much preferred hopping a plane into Dobo and out again after my days in the backcountry—but there aren't any planes into Dobo. The Aru Islands are still high on the list of most isolated places on Earth, and the sole way to visit them is the old-fashioned way: going and coming slowly, by boat. It's no wonder that Dinsmore and those other ornithologists contented themselves with Little Tobago. Emerging from the forest, I rejoined the cruise ship reluctantly and steeled myself for prolonged anticlimax. We'd be another ten days at sea.

Several nights later, while the ship lumbered in midocean, I woke with the frustrating realization that I had no physical token of my visit. I hadn't bought any keepsakes. I hadn't collected any specimens. I had made notes and gathered memories, but I hadn't even taken a photo. Now that Aru was behind me—and I would never see it again in this lifetime, I figured—I regretted going away empty-handed. Suddenly I wanted something I could touch. Damn it, damn it, I told myself: I should have bought those pearls for my wife.

A symptom of loneliness, no doubt. In the course of my work I travel too far and too often, for too many weeks at a stretch, from a woman I love dotingly and a home that's already my favorite remote place on the planet. I don't need these absences, thanks, to make my heart grow any fonder. But I try to mitigate them by spates of uxorial shopping. I get solace from saying, "Yes, I'll take this one, it's for my wife. The color is right, the shape is interesting, yes, I think she might like it." The point of such purchases is to forge a connection: her with the distant place, the place with her. Never mind that she happens to be a person of Doric simplicity of tastes, with small use for these trinkets and baubles I bring her. I continue buying them anyway, because they make me feel homebound and good: a woven dress from the Ecuadoran highlands, a silver bracelet from Kathmandu, a slab of ammonites from the Jurassic slates of Bavaria, a carved mask from Bali, a Huon-pine bowl from Tasmania, a fiber bag from New Guinea, a zebu-hide purse from Madagascar.

And from the current expedition . . . well, um, nothing. No souvenir, no memento, no acquisitions whatsoever. God knows, everyone was selling. But for some reason I hadn't been buying. Now I wished otherwise. It seemed that a trinket from Aru, destined for my wife, would have helped shrink the distance home. For an hour I lay in an insomniac sweat, regretting the Dobo pearls.

Then it occurred to me how the gentle biologist whose husband I am would feel. She'd have less desire for Aruese pearls than even for Nepalese silver. Quicker and more trenchantly than anyone else I know, she would sense that these little islands have already surrendered enough of their living and once-living treasures. This time I'll give her an essay, I decided, and together we'll let Aru be.

# *Bagpipes for Ed*

A man wrote a book, and lives were changed. That doesn't happen often.

Trees go to the pulp mills every day, paper gets made, presses roll constantly, and we bob along on a Noachian flood of printed words, a great tepid and oily deluge of discourse, nearly all of which is as dispensable as sewage. But this particular book was not dispensable. This book, by some miraculous convergence of honesty and insight and wit and good timing, struck firmly upon hearts and brains; it fastened. It mattered. Never mind that it had been published quietly in the noisy year 1968, with scant if any advertising support, and never mind that its author had no reputation. He was an obscure 41-year-old, the author in question—a former welfare caseworker, a former technical writer for General Electric, who had also done time in the employ of the National Park Service. His previous oeuvre consisted of three ignored novels. Now suddenly came a book that was different, more stark, a book full of landscape and meditation and voice. The landscape was graceless, by conventional standards, and the voice was as strong as garlic. It couldn't have looked like a promising package on Publishers' Row. Nevertheless people discovered this new, odd, cranky book; they found it on their own, by happy accident or hearsay, without benefit of publisher's hype or critical commendation, and they mailed paperback copies to each other with peremptory notes in the vein of: "Here. Trust me. Just *read* this thing." My brother-in-law Robert, back in 1973, mailed just such a packet to me. And so one of the changed lives was my own. I know of others. Possibly yours too? The book was *Desert Solitaire* and the author was Edward Abbey.

Now he's dead, though the book isn't. How should we think about that fact? How should we feel about it? How should we conduct ourselves in observance of this man's death?

Reverential bereavement can be ruled out, since Ed Abbey was himself the most determinedly irreverent of writers, of men, and the fumes of our piety would only wrinkle his nostrils. Anyway, there are better alternatives. Take a long hike in the desert. Go alone, for two or three days, and without a Walkman or a camera. Or take a long hike up a mountain, down into a canyon, over a prairie, through any piece of wild landscape you might be lucky enough to find still unspoiled—thanks partly to Ed Abbey—in this our much-despoiled country. Uproot the survey stakes as you go. Throw a beer can, metaphorical or otherwise, at some duly elected apologist for the holy inevitability of commercial and industrial growth. And of course: Read or reread Abbey's books. A dead author is always well honored by those folk who find his work undying.

In Abbey's case, you are offered a broad variety and a reasonable abundance of books (though Ed himself once disparaged it as appallingly small output for more than thirty years as a writer) through which to revisit his mind and to experience his unmistakable voice. Seven novels and a dozen volumes of miscellaneous nonfiction—essays, observations on life and landscape, revisionist memoirs, which he bridled at seeing labeled as "nature books" and preferred to call "personal history." The earliest was a novel called *Jonathan Troy* (1954) and the most recent (1988) another novel, *The* Fool's P*rogress.* There may also be a further novel, finished just before he died, to be published posthumously, possibly under the title *Hayduke Lives!** You haven't asked for advice so I'll offer some. Direct yourself first to *Desert Solitaire.*

*It was: *Hayduke Lives!* (New York: Little, Brown and Company, 1990.) I've left this essay grounded in the time of its composition, just after Ed's death in March 1989. For details of first publication, see "Notes and Provenance" at the back.

That's how I've been mourning Ed Abbey these past few days: by going back with him to the little trailer outside of Moab, back to Arches National Monument, back to the cliffrose and the juniper and the rattlesnake under the porch, the moon-eyed horse and the dead man at Grandview Point and the moment in a blind canyon near Havasu when our narrator barely escapes a terrifyingly lonely death. Back to "A Season in the Wilderness" (as the book's subtitle modestly bills it) during a fundamentally different era, an era before the Arches road had been paved and before Edward Abbey himself had become blessed and cursed with his status as America's most famous desert-loving curmudgeon. Back to that moment, fresh and transient as a desert dawn with snow, when Ed was just discovering his real literary destiny and a community of readers was just discovering him. *Desert Solitaire* begins on a first day of April (of a carefully unspecified year) and ends in October, and in between is encompassed a world and a life and a vision. If you've never read this American classic . . . why not?

If you have read it but not recently, why not again? My message to you is precisely the same as that one I received all those years ago from my brother-in-law: "Here. Trust me. Just *read* this thing." The book isn't merely good. It's better and more far-reaching than you remember.

*Desert Solitaire,* as I mentioned, was first published in 1968. If you recall that jangling and epochal year, or have heard tell, you know that America just then was preoccupied with matters far removed from the contemplation of cliffrose and juniper in southeastern Utah. No wonder the book wasn't an immediate best-seller. It was drastically, irredeemably out of step. Although Abbey had more elemental intentions than merely being unfashionable for contrariety's sake, still, contrariety was one of his great talents, and he never applied it to better effect than here. His desert journal was boldly irrelevant when published and therefore timeless ever after.

The closest he came to acknowledging the prevalent cultural clamor was in a few pharmacologically freighted metaphors:

> Noontime here is like a drug. The light is psychedelic, the dry electric air narcotic. To me the desert is stimulating,

> exciting, exacting; I feel no temptation to sleep or to relax into occult dreams but rather an opposite effect which sharpens and heightens vision, touch, hearing, taste and smell. Each stone, each plant, each grain of sand exists in and for itself with a clarity that is undimmed by any suggestion of a different realm.

The cliffrose and the juniper were a mantra conducive to meditation, though God knows Abbey wouldn't have said it in any such terms.

He surprised us with this book in a way that—surprise being what it is, a one-time fracture of expectations—he could never surprise us again. My view is that he also, to some degree, surprised himself. In one of the later books, he tells a story (as distinct from *the* story, since this account may contain more psychological truth than literal accuracy) about how *Desert Solitaire* came to be. Back in those years when he was still an obscure and not-anymore-so-young novelist, living in Hoboken of all places, an editor at a New York publishing house returned his latest manuscript with a letter, Ed claims, that said: "This alleged novel, Mr. Abbey, has no form, no content, no style, no point, no meaning. It's not even obscene—has no redeeming social value whatsoever. I advise you to burn it before it multiplies." Abbey, characteristically, rejected the rejection. He sent a stiff-backed reply that was so singular, so unassailably self-possessed, it caused the editor to invite him to lunch. A lunch invitation meant a free meal, so Abbey put on his galoshes and shuffled into Manhattan. The editor took him to Sardi's. That nameless editor, evidently a shrewd one, saw the chance of discovering a voice. He wanted a book from this Ed Abbey, but not a contrived fiction. The editor suggested, as editors routinely and unhelpfully do, that Ed write about something he knew. About Hoboken, possibly.

Ed ate the lunch and listened. But he didn't care to write about the urban wasteland. Instead: "I went back to my condemned tenement in Hoboken and typed up, out of nostalgia, an account of a couple of summers I had frittered away, playing the flute and reading Dreiser, in the Utah desert. *Desert Solecism*, I called the book—a title later corrupted to *Solitaire* by the publishers."

Swallow that one at your own risk. But do note Abbey's persuasive insistence on one point: that the creation of this book was somehow unpremeditated, offhanded, uncalculated. In writing it he surprised himself, I believe, with the lovely discovery that his powers of observation, his unadorned passions and convictions, most of all his singular voice, could be shaped into an act of literature just as potent—just as artful, even—as any assemblage of invented characters and plot.

Despite that lovely discovery, he always preferred thinking of himself and being thought of as a novelist. He seemed weary of the disproportional adulation that his first nonfiction book received. He might want to stub out his cigar in my eye socket now if he could hear me going on about that book. Nevertheless. *Desert Solitaire* is his best piece of work and one of the luminous triumphs in our language in this half of the century.

But what *is* it, exactly?

In the same sense that *Walden* is not merely or even chiefly a book about life in the woods (despite its subtitle), *Desert Solitaire* is not merely or even chiefly a nonfiction book about the desert. Nor is it just a book about wilderness and the West. Its essence is larger than that. It's a book about the power of landscape; about the rightness of human connectedness to landscape (potentially *any* landscape, including the streets of Hoboken); about a passion for liberty set within that connectedness. It's about being alone and doing alone. It tells us that life is incomplete without lusts and risks and contradictions. It teaches the important and hard-to-appreciate truths that rock is beautiful, that water is precious, and that the desert landscape of the American Southwest, spectacular as it may be, contains no precise substitute for Beethoven. As I said at the start, it changed some lives.

It's a book about how to live and how to die.

TOWARD the end of *Desert Solitaire,* Abbey recounts the story of the dead man at Grandview Point. This fellow was a sixty-year-old tourist who went missing, alone, without adequate water or savvy, in the desert not far from Arches. Abbey was recruited to the search party; and before long they found the man's body, bloated and flyblown after two days of desert heat. It lay in the tiny patch of shade under a juniper that was rooted in rock on the brink of a mesa, overlooking a precipitous drop to the White Rim and, still deeper, the gorge of the Colorado. It lay near the place called Grandview Point.

Abbey says of the man:

> He had good luck—I envy him the manner of his going: to die alone, on rock under sun at the brink of the unknown, like a wolf, like a great bird, seems to me very good fortune indeed. To die in the open, under the sky, far from the insolent interference of leech and priest, before this desert vastness opening like a window onto eternity—that surely was an overwhelming stroke of rare good luck.

This sort of statement might seem just facile attitudinizing, if the words came from a person who lacked the fanatical force of conviction to live and die in that spirit himself. Abbey, as it turns out, was entitled to the words.

How did Ed Abbey go? There's no mystery. The newspaper obituaries variously said "internal bleeding" or "circulatory disorder," and additional medical detail doesn't really add anything to our solace, our grief, or our comprehension. He had a recurrent problem with certain esophageal veins. His stubborn disposition and his contempt for the abstrusities of medical "life-support" technology were factors. Decades of hard living and a touch of sheer physiological bad luck were also involved. He was in and out of the hospital several times during the last couple months; finally he submitted to an operation, which might have saved him; but he was weakened, and he kept bleeding. Near the end, he announced that it was time for him to be taken out into the desert so he could die there, and once he was out there—wonderful, contrary bastard—he got better. Somewhat. But then again more bleeding. He died at home among family and friends. He died bravely and unselfishly, I've been told. "Anybody who knew him would have been proud of him," I've been told. He escaped the insolent interference of leech and priest.

Is he buried in the desert? Yes. Is it an unmarked grave? More or less. Was he laid to rest in something the same outlaw spirit by which he lived? Matter of fact, yes. Can we go there and pay homage? Of course not. What we can do is read him and leave his bones in peace.

He left a message. Actually he left an abundance of messages—

at least fourteen books full of them, not counting the last manuscript or the coffee-table collaborations—but one in particular that addressed the question of his mortal remains. It was directed to his wife, Clarke; she shared it with some of the people who were close to him; one of them, Doug Peacock, who may have been Abbey's most fiercely, recklessly loyal friend, shared it with me; and at some risk of invading privacy, but for reasons that go beyond idle curiosity, I share a bit of it with you. The message pertained to what Ed Abbey wanted done for him, and to him, after his death.

He wanted his body transported in the bed of a pickup truck. He wanted to be buried as soon as possible. He wanted no undertakers. No embalming, for Godsake. Disregard all state laws concerning burial. "I want my body to help fertilize the growth of a cactus," said the message.

As for graveside ceremony: He wanted gunfire, and a little music. "No speeches desired, though the deceased will not interfere if someone feels the urge. But keep it all simple and brief." And then a big happy raucous wake. He wanted more music, gay and lively music. He wanted bagpipes. "I want dancing! And a flood of beer and booze!" And a bonfire, said the message. And people making love. And meat, lots of meat. Beans and chilies. And corn on the cob. Only a man deeply in love with life and hopelessly soft on humanity would specify, from beyond the grave, that his mourners receive corn on the cob.

So on a Wednesday in late March, eight days after his death and some few days after his body had been set in its place, he got most if not all of what he had asked for. Two hundred people gathered in Saguaro National Monument, just over the mountains from Tucson, for a "celebration" of the late Ed Abbey. There were great tubs of a hot desert stew, concocted from meat of mysterious provenance ("poached slow elk," in the terms of this recipe, suggesting perhaps a privately owned cow assassinated while it was grazing on public land) by Doug Peacock. Another close friend, Jack Loeffler, blew taps on a trumpet. There were grief and booze and chilies. There were bagpipes. There was joy at the privilege of having known this man, at having heard or read his inimitable voice. Being overseas at the time, I wasn't lucky enough to attend the celebration myself. But, fourth week in March, north edge of the Sonoran desert—in my imagination I can see it, with the ocotillo and the saguaro in bloom.

I HAD received my own final message from Abbey just six weeks earlier. This is a small matter but also bears sharing, I think, for what it says about a rock-ribbed man who—though in some ways quite sensitive, though crackling with intelligence, though deeply reflective—believed in getting on briskly with the business at hand, whether that business be living or writing or maybe dying. I had recently sent him a card, having heard about his first hospitalization, back in January, and the close brush with death he'd had then. Take some care of yourself, I had said. America has only got just the one Ed Abbey, and we certainly can't afford to part with him.

He wrote back:

> *Dear David—*
>
> *Thanks for the card but there's nothing to get alarmed about. Got this little medical problem that kicks up every 4-5 years. Our friend Peacock tends to exaggerate a bit. Am finishing a book now, then going exploring around the Sea of Cortez in March. All's well. Best regards,*
>
> *Ed A.*

Never trust an anarchist.

He was sixty-two. He left behind a wife, five children, a father, more than a dozen pretty damn good books, and a masterpiece. We should all be so lucky. We should all live and die so well.

# *Point of Attachment*

Barnacles, to the undiscerning eye, are as boring as rivets. This is largely attributable to the erroneous impression that they don't go anywhere and don't do anything, ever. The truth of the matter is that they don't go anywhere and don't do anything merely sometimes—and that, other times, barnacle life is punctuated with adventurous travel, phantasmagorical transformations, valiant struggles, fateful decisions, and eating. The drama of barnaclehood has been underrated. But it's like baseball when the pitching is good, or grandmaster chess on TV: Proper appreciation demands patience and a depth of understanding. Barnacles aren't for everyone.

If you don't trust me on this, you can get a second opinion from the most eminent barnacologist in scientific history, who happens also to be the most eminent biologist: Charles Darwin. During the early phase of his career, after his round-the-world voyage on the *Beagle* but before he had published *The Origin of Species,* Darwin devoted eight years of eye-crossingly hard work to the description and classification of barnacles. Eventually he produced a definitive treatise, which won great (if narrow) respect from his scientific colleagues and was judged by later observers (as well as by Darwin himself, in hindsight) to be maybe the dumbest thing he ever did. Eight *years,* Chuck? For *barnacles*? While your epochal idea about natural selection was left simmering on a back burner? What were you *thinking*?

What Darwin was thinking, at least initially, was that he would spend just a few months describing a single peculiar species, brought back among his *Beagle* specimens from the coast of Chile. He began work in October 1846, with the intention of moving on to the write-up of his evolution theory (which had been ringing in his head, like a bell clapper, since 1838) as soon as this last little *Beagle*-related task was polished off. But he got carried away with the barnacles, and instead of making one verbal sketch he felt compelled to paint an entire gallery's worth of portraits, as well as to renovate the building in which they hung. He revised barnacle systematics so proficiently that modern barnacle biologists still cite him as the granddaddy expert. Decades later Darwin admitted in his autobiography that, although the barnacle project had some value, "I doubt whether the work was worth the consumption of so much time." His two volumes on the living species (as distinct from the fossil barnacles, which he treated separately) were published in 1851 and 1854. Bound together, they constitute *A Monograph on the Sub-Class Cirripedia*—more than a thousand pages long, unrelentingly thorough, intermittently interesting, graced with many fine engravings of barnacle anatomy, not available in stores.

The copy of *Cirripedia* that rests here on my own desk, like a literary cinder block, is on loan from a faraway library. It's a 1964 facsimile reprint in pristine condition, not having circulated much in the past thirty years, during which both Darwin and barnacles have been slightly out of vogue. I was inspired to consult it after a recent field trip along the shores of the Strait of Juan de Fuca, northwest of Seattle, where I spent long hours in good company, wet-footed, spying on the fauna of the intertidal zone. I hadn't gone there expecting to dote on barnacles—in fact, I was inclined toward the view that they're as boring as rivets—but once I had put my nose down into the barnacle world and left it there for a few hours, something about them struck me as oddly majestic.

My brother-in-law Steve, a software designer with an off-hours enthusiasm for the animals that live in tide pools, had organized this expedition. We were accompanied by the two venturesome sisters whose spouses we are (one of whom, my wife, was the only card-carrying marine biologist in the group, but she was gracious enough not to flaunt it among us galumphing amateurs) and a phalanx of fair-haired nieces. We timed our daily forays for when the tides were at their lowest. The stretches of littoral where we prowled were as rich as a rainforest. The sea urchins were abundant, the starfish were hefty, the sculpins were cute, the nudibranchs were colorful, the sea cucumbers were bizarrely tumescent as only sea cucumbers can be, the mussels reminded me about lunch, the chitons were gigantic and stinky; but what most captured my attention were the barnacles. They were extraordinarily calm. They made even the limpets look hyperactive.

The barnacles had presence. They were stalwart. They cowled their soft parts and their secrets. They survived in the highest intertidal zone and on wave-pummeled vertical walls, where other creatures couldn't. They were indifferent even to being walked on, a measure of aloofness that not many animals can claim. They were intricate, adamantine, fierce in their own way, and sublime. What they lacked in mobility, they made up for in resourcefulness. They put me in mind of a Zen koan—the paradoxical riddle with no rational answer, by which the Zen master teases the student toward a level of enlightenment that transcends rationality.

Master asks student: "Why a barnacle?" Student spends five years at the seashore with furrowed brow.

That's how I felt. The barnacles of Juan de Fuca went nowhere and did nothing, during those hours of low tide, and I couldn't stop wondering what they were up to. Darwin himself had given these beasts almost a decade of his life, I recalled. I began to suspect that he'd had good reason.

THE NAME *Cirripedia* derives from *cirri,* meaning slender appendages. It might apply to tentacles, barbels, antennae, or almost anything else that sticks out. The second half of the name, *pedia,* is a hint that barnacles are not exactly what they seem: Despite the hunkering, lumpish appearance, they do have legs. But you've got to know when to look, and how to see. Among barnacle larvae, which are mobile and planktonic, the legs serve for swimming. With metamorphosis to adulthood, they become reshaped into food-gathering fronds.

Another counterintuitive fact is that barnacles aren't mollusks, despite the open-and-shut mechanism of their calcareous shells. Nor are they closely related to other lumpish marine animals, such as sea urchins, sea anemones, or sponges. Instead they belong to the subphylum Crustacea, along with lobsters and crabs. T.H. Huxley, a colleague of Darwin with an un-Darwinian gift for vivid expression, explained that a barnacle is a crustacean cemented head-down to its substrate, kicking food into its mouth with its legs.

The sex life of barnacles is wonderfully lurid. The typical species is hermaphroditic—equipped with both male and female organs. Some of those hermaphrodites are capable of self-fertilization, useful in lonely circumstances, but among most shallow-water species the rule is cross-fertilization, which is less convenient but presumably more fun, and advantageous on genetic grounds. Cross-fertilizing species don't simply spew out their eggs and their sperm into the water and then rely on external fertilization among the spawn of other spewers, as oysters do. They actually copulate, one barnacle reaching out to another with a prodigiously long penis. At this point it's appropriate to address a question that we've all probably asked ourselves at one time or another: Just how long *is* a barnacle's penis?

"Its length varies much, according to its state of contraction or relaxation," Darwin revealed. It's also endowed with transverse and longitudinal muscles, he added, which make it "capable of the most varied movements." In a small specimen of the barnacle *Elminius modestus,* Darwin's own patient dissection had turned up a penis that was three times as long as the animal's thorax. More recently, a survey of barnacle biology in the *Encyclopaedia Britannica* asserts that barnacles of some species, acting as males, "are able to inject spermatozoa into the mantle cavity of an individual as far as seven shell diameters away." In *Octomeris angulosa,* by contrast, Darwin found that the penis was quite short. A short penis is no laughing matter if you happen to be a cross-fertilizing barnacle that has cemented itself to a rock and then discovered that your nearest potential mate has cemented her/himself nine or ten inches away. You could sprain a longitudinal muscle while trying to establish meaningful relations.

Barnacle eggs hatch inside the mantle cavity of the female-acting parent, from which the young are expelled for their circuit through the marine environment at large. The first stage of larval development is a *nauplius,* a tiny six-legged form with one eye, fairly similar to the nauplian larvae of other crustaceans. The nauplius swims weakly and eats phytoplankton. After a couple weeks of growth and molting, it transmogrifies into a *cypris,* another larval stage, which bears little resemblance to either the nauplius or the adult. A classic of the marine literature titled *Between Pacific Tides,* by Ed Ricketts and a string of coauthors, describes the barnacle cypris as "an animal with three eyes, two shells, six pairs of legs, and an inclination to give up the roving habits of its youth and settle down." This is loose talk for scientists, since it smacks of anthropomorphism, which science officially abhors; yet it states the crux of the matter perfectly. Ricketts, John Steinbeck's old pal from Cannery Row in Monterey, knew and loved marine life and apparently didn't worry himself much over the protocols of professional rhetoric.

The cyprid larva swims strongly but doesn't eat. Its particular role within the full life-history pageant is to make a single consequential decision and perform a single irreversible act: It chooses a point of attachment—on some rock face or wharf piling or buoy, or on the flank of a whale, the hull of a ship, the shell of a turtle, the stem of a piece of kelp—and it cements itself to that point for what will be the rest of its barnacle life. Then it metamorphoses again, this time into a sessile adult.

The word *sessile,* which in its zoological application means "permanently attached," encompasses a large share of the essence of barnacles—what distinguishes them not just from restless terrestrial fauna, such as jack rabbits and cheetahs and tortoises and sloths, but also from most other animals of the intertidal zone. Chitons and limpets and starfish aren't sessile; they're merely slow. A mussel isn't sessile; it can schlep itself around with its single muscular foot. A cockle is acrobatic compared to a barnacle, which is genuinely sessile.

The force of this distinction became manifest after my youngest niece picked up a purple sea urchin out of a tide pool.

Having admired the long spines and the stony, globular body, she set it gingerly back into the water. The urchin, now riled, or dissatisfied with its new position, began to creep across the bottom, like an ottoman edging its way toward the kitchen. A sea urchin has feet, I remembered, not just for feeding itself but for walking. A barnacle, after adolescence, can only dream (or reminisce) of such mobility. It has many useful capacities but not the capacity to move—not ever again, not even a little.

During its few weeks as a planktonic larva, the barnacle may have traveled five hundred miles. During its years of adulthood, it won't travel five inches. It will make do instead on the basis of what the tides bring, what it can reach, and what it can endure. In the evolutionary contest, its prospects will depend on the duration of its survival, the fecundity of its ovaries, and the length of its penis.

BUT NOT every cypris graduates to adulthood. The great majority don't. For each cyprid larva that establishes itself and grows into a robust adult, untold numbers cement themselves into disastrously inhospitable circumstances and then die prematurely—from predation, competition, starvation, desiccation, sunstroke, and all the other sorts of dire hardship by which natural selection (as Darwin explained, when he finally got around to it) adjudicates evolutionary success and failure.

Most of the barnacle species fall within two major types, similar in their larval stages but visibly different as adults. The goose barnacles (Lepadomorpha to the experts, treated in volume one of Darwin's opus) stand on fleshy peduncles, resembling little gosling heads on flexible necks. Their calcareous plates are fitted together like inset tiles with flesh between, not fused into a single domed shell. The acorn barnacles (Balanomorpha, treated in volume two) are fully armored, cemented flush to the substrate, and nubby enough to inspire that analogy with rivets. They're the type with which most people are vaguely familiar. Along the Juan de Fuca coastline, we saw both: stunningly red-lipped goose barnacles of the species *Pollicipes polymerus* in the lower intertidal zone, and a mélange of acorn barnacles, such as *Balanus glandula*, farther up.

The ecological dynamics that underlie patterns of barnacle distribution are complex. As a group, the Cirripedia cover a vast breadth of habitats, from the Arctic to Cape Horn and from deep water to almost the level of the highest high tide, but the habitat parameters of any single species are more narrow. For instance, some acorn barnacles favor points of attachment along the upper fringe of the littoral, well above the range of most other intertidal animals. Their tightly closed shells allow them to endure long periods of desiccation and direct sunlight between high tides. But staying shut for so long also subjects them to food deprivation, since they depend on the back-and-forth slosh of wave action to bring them their meals. A closed barnacle is not a complacent barnacle if it's getting godawful hungry. Settling at the extreme upper fringe of the intertidal zone, on the other hand, allows an acorn barnacle like *B. glandula* to put some distance between itself and a whole roster of marine predators (notably the big orange starfish *Pisaster ochraceus* and the shell-drilling snail *Nucella emarginata*) that don't share its tolerance for exposure. Ecological studies have suggested that the upper limit of acceptable habitat for such species is set mainly by physical factors (desiccation, solar radiation) while the lower limit is set mainly by biological factors (predation, competition). In between, *B. glandula* occupies a narrow band, sort of an ecological DMZ, along which the two categories of otherwise mortal travail are both merely venial.

One laboratory study of certain *Balanus* species has revealed that instinctual gregariousness helps determine where any single cypris might settle. The larva at that stage seems to be capable of recognizing other members of its own species when it bumps into them. It lands on a plausible surface and walks around briefly, using its antennae to explore the terrain. If nothing catches its fancy, it may swim off to explore elsewhere. If it encounters a barnacle of its own species, though, the cypris more likely decides to stay, cementing itself in place and beginning its metamorphosis to adulthood. The experimental report on this phenomenon, by two scientists in North Wales, was published in *Nature* forty years ago. But just how a cypris recognizes its own kind (by touch? by taste? by some barnacle mating call too subtle for human ears?) seems to be one of the still-unanswered questions of barnacle ecology.

Then again, maybe not many people have asked.

* * *

DARWIN himself was especially struck by the cementing procedure, which he called "the most curious point in the natural history of the Cirripedia." From the evidence of his own dissections, he argued that the cement gland at the base of each antenna was a modified extension of an ovarian tube and that the animal used ovarian secretions to cement itself permanently to its substrate. Of course this modification of old organs for newer purposes implied an evolutionary process—but Darwin made no explicit claims about evolution in his analysis of barnacle anatomy. He was saving that subject for later.

Why was he saving it for later? Why did Darwin divert himself for eight years with barnacles, while the preeminent insight of his career—and of all modern biology—remained secret, roughed out in his notebooks and several sketchy manuscripts, confided only to a few trusted colleagues? Darwin biographers and other science historians have grappled with that mystery. Michael Ghiselin devoted a whole chapter to it in *The Triumph of the Darwinian Method.* Stephen Jay Gould, in one of his earliest popular essays, labeled it "Darwin's Delay." The prevailing explanation has been that Darwin earned himself a crucial sort of credibility through this long, tedious exercise in systematic zoology, and that he had foreseen that he would need a big share of such credibility when the time came to announce his outrageous theory of evolution. His closest scientific friend, Joseph Hooker, had argued to the effect that "no one has hardly a right to examine the question of species who has not minutely described many," and Darwin took the warning to heart. Gould, a careful but sometimes cranky scholar, dismissed that explanation as "pap" and proposed another: fear. Darwin was acutely aware that his theory would cause some ferocious controversy, according to Gould, and his queasy-stomached dread of such controversy played a large role in creating the two-decade lag between conception and publication of the theory—two decades that included the eight years spent fussing over *A Monograph on the Sub-Class Cirripedia.* In other words, the barnacle project was displacement activity. Darwin was stalling, because he didn't want to set Victorian piety on its head.

Both explanations have their merits, but they both tend to

neglect an additional pair of factors. The first is that barnacles are inherently fascinating. They damn well do happen to be complicated and significant enough to enrich eight years in the life of a serious biologist. The second factor is that the life history of barnacles bears a totemic resemblance to the life history of Charles Darwin himself. My own little theory about Darwin's cirripedological delay—offered merely as a supplement to the other two—is that he became transfixed by a deep-seated empathy for these animals, almost as though he were gaping into a mirror.

Let's review a few facts and dates. After his five years of circling the world on the *Beagle,* during much of which he was seasick and homesick, Darwin set foot back in England in 1836. In 1837 he began his first notebook on the question of "transmutation" of species, and in 1838 he hit on his great idea: that natural selection was the mechanism by which it happened. Meanwhile he had taken up residence in London, which he hated as an "odious dirty smoky town" but which allowed him access to museums, libraries, scientific societies, and the dinner-table company of brilliant people. In 1839 he married Emma Wedgwood, his first cousin, who was not the dinner-table-brilliant type. Then, in 1842, he bought an old parsonage in the hamlet of Down, "a quiet most rustic spot," two hours by carriage southeast of London. He and Emma moved there in September. For the next forty years of his life, which was all of it, Down House (as it came to be called) was the only home he knew or wanted.

He'd had a bellyful of travel. His health was bad, he was shy, he was busy, he was defensive and secretive about his most important ideas; for these reasons and who knows what others, Darwin became an obdurate homebody. On his lonesome he wrote books and on Emma he fathered a passel of children. He never again went abroad. He traveled around Britain occasionally, but not far, not often, and only for extraordinary personal or professional reasons, such as taking the water cure at a provincial spa. Mostly he stayed at Down, reading, writing, dissecting barnacles, running little experiments in his study, being a good husband and father. He dropped away from the scientific societies of London and made clear to his close colleagues that if they wanted to talk, they would have to come visit at Down. He did his more far-flung research through the mail, pestering scientific pen pals all over the world for bits of data. When his father died, in 1848, he missed the funeral. When his eldest and much-beloved daughter died, in 1851, he skipped the funeral. Both these funerals, you'll notice, fell within the barnacle years. When the British Association held its famous meeting at Oxford—in 1860, a year after *The Origin of Species* had triggered an intellectual revolution—and T.H. Huxley debated Bishop Wilberforce about the new theory's profound and scandalous implications, Darwin was conspicuously absent. Again he preferred to take an invalid's excuse and let others report to him what had happened. No doubt he'd have skipped his *own* funeral, at Westminster Abbey in 1882, if allowed any choice in the matter.

Having traveled so many hard miles in his young manhood, Darwin had found his preferred spot—that old house in the hamlet of Down—and cemented himself to it. Virtually everything he wanted or needed (except quack doctors offering water cures) was there within reach. So why move? Small wonder, if you ask me, that he felt a special attachment to the Cirripedia.

As THE Pacific bulge pushed itself back up into the Strait of Juan de Fuca and the rising tide nudged me toward high ground, the acorn barnacles of the uppermost littoral were the last creatures over which I could ruminate. I was intrigued by the way each one lives out the consequences of that single decision, made as a larva: where to attach. How do they choose? How do they know? How do they achieve such heroic resoluteness? Wet to the shins, I was still listening for the call of the barnacles.

But no, nothing. Only the susurrus of sea against land.

Maybe Darwin had heard it, I thought. Maybe he was patient and focused enough to apprehend their most private truths. I made a mental note to get hold of his big, boring, authoritative book, the one some people felt he shouldn't have bothered to write. Then we hopped in the car and wandered on.

# *Voice Part for a Duet*

Two intriguing statistics recently grabbed my attention. They concern that remarkable form of social behavior known as monogamy. The first is a testament to avian fidelity: Ninety-two percent of all bird species are monogamous, at least through a single breeding season. The second is another sort of testament: Among *mammal* species, in contrast, just three percent are monogamous. By my clumsy arithmetic, that makes monogamy about thirty times more prevalent in the avian world than down here on solid ground among the class of plodding beasts to which you and I and Elizabeth Taylor belong. This of course leads to a question about mammals versus birds: Who are we to call *them* flighty?

Another question carries more scientific import: Why is mammalian monogamy so rare? If there's a telling pattern within the data set, it doesn't show at first glance; the monogamous mammals seem crazily varied. The trumpet-eared bat is among them, and so is the beaver. Also the bowhead whale. A fair number of canid (dog family) species are monogamous—the wolf, the coyote, the golden jackal, the red fox, the arctic fox, the Cape hunting dog, and others—but no felid (cat family) species, no bears, no hyenas, and just one marsupial. The klipspringer, a dainty African antelope, is monogamous. So is the indri, the largest of the lemurs, an extraordinary animal whose ghostly yodeling haunts the rainforest of northeastern Madagascar. Most other mammals are not. Monogamy, then, is like flight: an exceptional phenomenon among mammals, a sort of defiance of gravity by alternate means. It occurs in certain species but not many, and the circumstances have got to be special.

What accounts for the low incidence of mammalian monogamy? Well, don't jump to the wrong conclusion when I say that some biologists attribute it to the invention of the breast.

IN A PAPER published some years ago in the *Journal of Theoretical Biology,* Allen T. Rutberg wrote: "The possession of mammary glands by female mammals encourages heavy parental care by females and biases mammalian mating systems in favor of polygyny." Rutberg's *polygyny* is the precise biological term for one male's matings with multiple females. Birds don't experience the same sort of physiological encouragement, since a male bird can provide parental care—incubating an egg or dropping worms into the maw of a hatchling—in essentially the same ways as a female. The most vivid exemplar of that truth is the male of the emperor penguin, *Aptenodytes forsteri.* This steadfast papa spends two months in midwinter standing deserted, on the Antarctic ice, with his mate's single egg balanced on the tops of his feet. He keeps the egg warm there, beneath his abdominal flap, while the female goes off on a feeding excursion to build her weight back up. When she returns, finally, he gets his own turn for an overdue meal. His reward for paternal reliability is higher reproductive success than he could expect if he mated with many females and abandoned them all. If he chose the latter strategy instead, he might fertilize six or eight eggs from six or eight different mothers, but probably not a single offspring would survive. Life on the ice is so hard that it takes two emperor penguins, cooperating from start to finish, pulling long shifts, to rear one chick. The male can do his full share because breasts don't enter into it.

Among mammals, the canids aren't unique in their inclination toward monogamy. The primates lean that way too—not overwhelmingly, but disproportionally relative to the overall ratio for mammals—with about fifteen percent of their species monogamous. That figure includes the indri but no other lemurs. It might or might not also include *Homo sapiens,* since humans can be seen as either monogamous or polygamous, depending on which human culture you look at. According to one ethnographic source, monogamy is the standard marital arrangement in just 137 out of 849 societies examined. So human monogamy, such as it is, stands as an odds-against exception (137/849 being the ratio of societies) within an odds-against exception (15/100 being the ratio of primate species) within an odds-against exception (3/100 being the ratio of mammal species) to prevailing patterns. Never mind the arithmetical bottom line. The point is that monogamy—among mammals in general and humans in particular—is an anomalous sort of behavior that can't be taken for granted.

Why *do* the evolutionary odds seem to set themselves so strongly against monogamous behavior among mammals? Is there more to it than the matter of breasts? And why have some species, breasts and all, flouted those odds?

Unfashionable thoughts: Monogamy is daring and mysterious.

BIOLOGISTS have always been interested in mating habits—and who can blame them, since the subject holds great prurient appeal as well as scientific significance. But for the past thirty-some years there has been increasing speculation about the origins of monogamy versus polygamy. Back in 1964, for instance, a researcher named Jared Verner published a paper titled "Evolution of Polygamy in the Long-Billed Marsh Wren," which helped launch the discourse into its modern phase. The ornithologist David Lack continued it in his 1968 book, *Ecological Adaptations for Breeding in Birds,* contributing some semantic clarity by dividing the concept of polygamy into harem polygyny (one male monopolizing several mates for a period of time), successive polygyny (one male mating successively with several females), and polyandry (one female mating with several males). Lack also wrote of the pair-bond, an implicit behavioral contract that ramifies beyond the act of copulation itself and distinguishes polygyny and polyandry from sheer promiscuity. The pair-bond can be limited to a brief phase of consortship or to a single breeding season, though in some species (the Canada goose, the klipspringer, the indri) it lasts for a lifetime. About polygyny, polyandry, and promiscuity, Lack said something notable: "The ecological factors making these abnormal types of pairing more advantageous than monogamy are not clear." The notable part is that he called them "abnormal." Later biologists who focused on mammals more than on birds would reject Lack's assumption that monogamy was normal.

Edward O. Wilson, having come to theoretical ecology and sociobiology by way of the study of ants, also rejected that assumption. In his compendious 1975 volume, *Sociobiology*, Wilson said that polygyny would be the natural arrangement for any species in which female sex cells (eggs) are larger and more expensive metabolically than male sex cells (sperm). Producing those big, nutrient-rich eggs requires the female animal to invest more bodily resources in each single act of mating than the male, who has little to lose by mating frequently. Wilson suggested that monogamy, on the other hand, might evolve secondarily in response to any of three ecological circumstances: (1) some sort of crucial habitat resource (a good nesting hole, say) is so scarce and localized that the cooperative effort of two adults is required to defend that resource against competition; (2) the physical environment is so difficult (as it is for the emperor penguin) that cooperative effort is required to cope with it; and (3) early breeding (at the first flush of spring, perhaps) is so advantageous that monogamy saves precious time that would otherwise be squandered on courtship. Wilson's third condition is relevant to the evolution of long-term pair-bonding. Finding a new mate each year might be amusing but it's also costly, and lifelong monogamy—at least for some species—is a more economical reproductive strategy.

The zoologist Devra G. Kleiman published a hefty review article, "Monogamy in Mammals," in 1977. Kleiman's piece is especially interesting for its attention to the social dimension, as well as the sexual dimension, of mating. She recognized that certain forms of gentle, sedate behavior—and not just the hot immediacy of courtship and copulation—play an important role among monogamous species. The simple act of resting together, for example, occurs conspicuously among beavers, among the antelopes known as dik-diks, and also among such monogamous primates as the marmosets, the tamarins, the siamang, and the gibbons. Another sort of amiable sociality is grooming—that is, picking carefully through another animal's fur to remove parasites—which is common among primates generally but shows itself in a special way among monogamous primates: Male partners more often perform the grooming on females, a nice little token of the pair-bond. Kleiman also mentioned the monogamous titi monkeys of the genus *Callicebus,* of which mated pairs sometimes sit side by side with their tails twined together. Tail-twining doesn't produce any new offspring or directly aid the survival of those already born, it doesn't even get rid of parasites, but it may well serve some adaptive purpose in the long run. It's another reaffirmation of the pair-bond.

Kleiman was intrepid enough to discuss the pair-bond between monogamous humans. She noted drily that, in Western society, "great emphasis is currently placed on maintaining high levels of sexual interactions in married couples" after they have started raising children. It's the old how-many-times-a-week? question that young husbands and wives begin asking themselves, proudly or disconsolately, around their fifth anniversary. "Frequent sexual behavior is thought to contribute to the maintenance of a strong bond in humans," Kleiman wrote, whereas "clearly this is not the case in many species of monogamous mammals where contact and affiliative behaviors, such as resting together and grooming, are more common than sexual behavior." She and her titi monkeys were onto something.

In the two decades since Kleiman's article, others have speculated further about the evolution of monogamy. Their ideas are multifarious. A pair of researchers named James F. Wittenberger and Ronald L. Tilson, in a 1980 paper, described three preconditions that must exist before monogamy can evolve and five hypotheses that might account for just why it does evolve when it does. They judged each hypothesis against empirical evidence from various kinds of animals—colonial birds, ducks, carnivorous mammals, antelopes, primates, reptiles, frogs and toads, wood roaches, dung beetles, horned beetles. Monogamy among the horned beetle *Typhoeus typhoeus,* you'll be excited to learn, is attributable to hypothesis three, the one about individual males stashing individual females away so that no other males can get to them. Details and tape at 11:00.

Allen T. Rutberg, whose observation about mammary glands I quoted earlier, divided the question of monogamy into two processes of choice—female choice about how to distribute themselves on the landscape, male choice about how to react to the females' distribution—and then analyzed the factors that can influence each choice. C.P. Van Schaik and R.I.M. Dunbar listed four hypotheses, three of which were reshaped versions of what Wittenberger and Tilson had proposed, one of which was more fresh, and from each hypothesis generated a handful of testable predictions. Van Schaik and Dunbar's paper was titled "The Evolution of Monogamy in Large Primates: A New Hypothesis and Some Crucial Tests." The two authors matched their predictions against known facts about the behavior of gibbons, indri, baboons, humans, and other primates.

What does all this theoretical work tell us? It tells us that monogamy isn't one behavioral pattern but many, and that it might or might not arise for different reasons in different situations. The scientific literature on the evolution of monogamy is intricate, ingenious stuff. You could read your way through a fair portion of it, as I've lately done, and be almost assured of replacing your ignorance with confusion.

BUT MAYBE I can minimize your confusion, and relieve my own, by lumping all the hypotheses and preconditions and predictions together and rendering them down to just a few simplified points. First, monogamy is more likely if females of the species spread themselves thinly across the landscape than if they gather together in big sisterly aggregations. Why? Because if the females are far enough apart, a male will run himself ragged trying to maintain a harem. Second, monogamy is more likely if males of the species have the physical or behavioral capability of making some crucial contribution toward the gestation and the rearing of offspring. That contribution might be direct, as in male birds or canids who carry food back to their young, or indirect, as in the male klipspringer, who stands lookout for predators while the female eats. It might involve the male of a given species defending a territory of rich habitat, in which his mate enjoys the right to feed without competition from other hungry females or interruption by randy males. It might even take the form of the male protecting his offspring from infanticide by rival males, murderous interlopers who covet his mate, his territory, or both. If the circumstances are right, the trend toward male contribution will be promoted by natural selection, as males who do contribute leave more offspring than males who don't. And the logical extreme of the trend is an exclusive helpmate relationship with one female.

Here's a third point: Monogamy is disproportionally common among primates because male parental investment is both more possible and more crucial than it is among other mammals. Remember, primates have dexterous hands and big brains. The big brains grow slowly and entail longer periods of juvenile dependency during which noninstinctive behaviors must be learned — and longer dependency requires more parental investment. The dexterous hands allow males to contribute in a variety of ways, despite their pathetic, embarrassing lack of breasts. A male zebra just isn't capable of making himself as useful, paternally, as a male marmoset who carries his young through the treetops. So the zebra has no better option than to mate with multiple females and hope that *some* of his progeny will survive. The marmoset plays a different gamble.

Marmosets are small-bodied animals, and so are most of the other primates among which males carry the young or bring them food. Large-bodied monogamous primates, like the nine species of gibbon or the indri, generally don't provide that sort of direct paternal contribution. But they do form durable pair-bonds, which are manifested in various ways: traveling with the mate, sleeping beside her, grooming her, performing soulful singing duets with her. According to a tabulation by Van Schaik and Dunbar, seven of the nine gibbon species, as well as the indri and the Mentawei leaf monkey, engage in duetting. The duetting seems to be a means of reaffirming to each other, and announcing to the wider world of predators and competitors, an irrefrangible mutual commitment. What's the ultimate impetus for that commitment? Van Schaik and Dunbar offered a single answer, the "new hypothesis" of their paper's title, persuasively supported with logic and evidence: Among larger primates, pair-bonding prevents infanticide.

Like Kleiman before them, Van Schaik and Dunbar also ventured into the subject of human pair-bonding. Yes, the new hypothesis is applicable to *Homo sapiens,* they argued: If human babies didn't face a high risk of fatal abuse, human adults might not be inclined toward monogamy. In support of this notion, Van Schaik and Dunbar cited two chilling sets of data. One study, done in Canada, indicated that stepchildren are sixty-five times more likely to die before their second birthday than children living with both biological parents. Another study, among the native Ache people of Paraguay, found a similar pattern: Children whose fathers died or departed were fifteen times more likely to die as youngsters themselves. It makes monogamy seem mortally pragmatic.

BUT MONOGAMY isn't mortally pragmatic. Not in all cases, anyway. If it were, we could understand it better. Just as surely as the weird, ethereal howls of an indri duet contain some transcendent dimension of grace—and they do, I promise you—so does the weird, narrow mystery of monogamy transcend Darwinian explanation. The data on which I base that assertion are less formal and orderly than Van Schaik and Dunbar's, but still persuasive: The world is filled (though sparsely) with voluntarily childless human couples whose bondings endure, even thrive, against all the social and biological odds.

Some people like to argue that childless couples are selfish. On a planet already worn threadbare beneath its burden of human population, that notion seems too perversely delusional to merit refuting. Anyway, I'm not going to bother refuting it, because the subject at hand is monogamy, not ecological overshoot. The subject at hand is tail-twining and song.

I can imagine a case of human monogamy in which all the hypotheses of the scientists don't apply. I can imagine a case in which the gestation and rearing of young doesn't figure and there's no egg to be held on the tops of the male's feet through a long Antarctic winter. I can imagine a case in which habitat resources aren't scarce, in which the physical environment isn't too difficult, in which early breeding at the first flush of spring isn't the least bit advantageous. A case in which infanticide presents no concern. I can imagine a case, say, in which the man spends a month in Madagascar or somewhere almost as remote and, with that absence, his heart grows fonder for no pragmatic Darwinian reason. I can even imagine that, when he hears duetting indri on a hillside in the northeastern forest, their glorious tangle of voices evokes in him a twinge of envious melancholy, as well as thrilled admiration. I can imagine that this sort of linkage between one human male and one human female might evolve slowly over a period of years, from the hot immediacy of youth to the sedate forms of affiliative behavior that become more important in middle age, and beyond. I'm talking now about that part of the social dimension even sociobiology can't explain.

Biologists call it pair-bonding. Others simply call it love. Let's us call it true and lasting love, as distinct from more familiar varieties. It's one of those unfunctional and gratuitous phenomena, like the sonnets of Shakespeare, the paintings of Mark Rothko, the music of Andrés Segovia, that make humanity just a little peculiar.

# *Love in the Age of Relativity*

Fourteen people sit gathered together in a room, performing a ceremonial measurement. Here, to this point, they declare, marks fifty years. For the matter at hand, a marriage, fifty is the roundest of figures. Pearl Harbor hadn't yet happened. Henry Fonda was young. The wedding took place in a world without television, without the Bomb, a world of Elgin and Bulova watches with jeweled movements but a world lacking any such thing as a Timex digital chronograph. Then the world changed but a few things remained constant, and lo, fifty years later this marriage is one of them. Someone speaks a toast. Goblets go up, wine goes down. In the air is a flavor of joy, gratitude, triumph against the winds of modernity, love, and a little quiet amazement. The number fifty seems both grave and festive. It seems stupendous. It fills the mind with all sorts of human questions, and with one that can be taken as scientific: What do we measure when we measure time?

Picture this scene with me. The room is a private banquet hall on the second floor of a fancy hotel, downtown in a certain midwestern city. Windows overlook a civic plaza, with its statuesque fountain, its trees, branches cheerfully strung with small yellow lights. The planes have arrived at the midwestern airport, roughly as scheduled, and the kin have assembled—daughters, a son, their spouses, grandchildren. Also a handful of old friends. Fourteen is a low number of place settings to encompass so many categories of people and the atmosphere is expansive but intimate. Dinner materializes from backstage. Ceviche, chicken and salmon, perigourdine sauce whatever that is. A bit more wine. Then a small woman with Irish red hair, the bride of fifty years' standing, pulls the gift-wrapping off a large box. She untangles padding. She lifts out an elaborate clock, crystal and gold and polished hardwood beneath a glass bell. She tilts her head back and laughs hilariously, because the clock comes as punch line to a joke.

It's an inside joke, an unintentional joke derived from improbable circumstance, and it will require a touch of explaining. We'll get to that explanation in due time. Meanwhile, the moment itself. This particular tableau on this particular evening isn't captured by any flashing Minolta. But memory can be lithographic. The small red-haired woman laughs hilariously.

Her daughters were smarter than her son, who doubted briefly the wisdom and good taste of choosing a clock to commemorate a fiftieth anniversary. Her son was wary of what he took for the double-edged symbolism. Anyone who has been married for fifty years—thought the son—and who therefore has passed seventy-some-odd birthdays, doesn't need to or want to be reminded about the inexorable tick-tock of time. It seemed faintly morbid. But what did he know? Stop thinking so hard, said his sisters. He was happy to be overruled by their wiser heads. The clock chosen is a handsome mixture of high-tech precision and old-world craft, from Kern & Söhne of Germany, with a quartz controller and a crystal pendulum that twirls rather than swings. The twirling, he admitted, is a nicely nonlinear touch. Now it sits, golden and unawakened, on the banquet table. Batteries not included. Time is momentarily frozen, according to this discreet gift. The groom of fifty years standing wears a contained but expressive Norwegian smile. Irish laughter fills the room like chimes.

WE LIVE in an age of shattered absolutes, and one of those absolutes is time. It's not what it used to be. For this we can blame

Albert Einstein, among others. Just as the Newtonian revolution in physics (with its own sort of action-and-reaction relativity) had destroyed the idea of absolute space, absolute stillness, absolute movement, Einstein's revolution destroyed the idea of absolute time. Space-time, a four-dimensional matrix, replaced the three dimensions of space and the linear continuum of time. Simultaneity became a dubious proposition, something that could only be claimed for two events that occurred in the same spot. The question *when?* became inextricably linked to the question *how far away?* The special theory of relativity decreed this rejection of time's absoluteness in order to make sense of the speed of light, which was itself an oddly relativistic absolute.

A startling experiment by two physicists named Michelson and Morley had shown that light always seems to move at one speed (186,000 miles per second), whether or not its source is moving also. That violates Newtonian dynamics, in which a spear thrown at forty miles per hour from a horse galloping at thirty miles per hour will hit its target at a summed speed of seventy. The speed of light, Michelson and Morley discovered, doesn't sum. To a hypothetical astronaut, moving at half of light's speed himself, the headlight beam on his spaceship seems to flash forward at 186,000 miles per second. To a ground controller on the planet of destination, the beam also seems to arrive at just 186,000 miles per second—not at that speed plus the speed of the spaceship. The rectification of this paradox is achieved by recognizing time as relative.

Put simply: The astronaut and the ground controller measure light's speed with two different clocks. The figure of 186,000 miles per second remains constant, between the two witnesses, because the duration of a second does not. The astronaut's second is much longer than the ground-controller's. With his spaceship moving so quickly, his clock has slowed down. This is one of the tenets of relativity: Speed slows time. Gravity also slows time. A clock on a huge massive planet, like Jupiter, will advance more slowly than the same clock when it's on Earth.

"What is a clock?" Einstein asked, in a surprisingly readable book called *The Evolution of Physics.* "By the use of a clock the time concept becomes objective." But *objective,* as he meant the word here, should not be confused with absolute. The invention of clocks merely objectified the flow of time into countable increments. Whether those increments were inherent or arbitrary, constant or changeable, cosmically truthful or imaginary, is another question. "Any physical phenomenon may be used as a clock, provided it can be exactly repeated as many times as desired," Einstein added. For instance, a heartbeat might serve as a clock. Geese flying south might serve as a clock. An oak tree—budding, leafing, turning crimson, going bare, budding again, leafing again—makes a rather good clock, highly reliable though not precise. Dripping water and falling sand were harnessed in clock-like devices and performed fairly well (except when the water was frozen or the sand was wet), until mechanical clocks became popular in the thirteenth century. But the oldest and most dependable clocks were the sun and the moon and the stars.

Celestial timekeeping has been universal to all human cultures. Sometimes it was elaborate, as with the Mayans, and sometimes rudimentary, but always there. The rising and setting sun measures a day; the sun's movement around the horizon, or the stars' movement around the night sky, measures a year; the waxing and waning moon measures a month. Unfortunately, none of these measurements quite agrees with the others, so time as marked off by our species has always been a confused set of numbers and labels with a margin of error that continually grows larger, and so requires continual adjustment, like an off-balance Maytag in need of a shim. By the moon's cycle, a month is 29.53 days long. By the sun's cycle, a year is 365.24 days long. How long is a day? Using the stars as referent, it comes to 23 hours, 56 minutes, and 4.1 seconds; but using the sun, it comes to 24 hours, 3 minutes, and 56.55 seconds. The cycles are incommensurable, an astronomer would say. The numbers don't match, either directly or as multiples. No wonder we've had to patch the system with leap years. No wonder we've got months of four different sizes, arranged so illogically that they can only be remembered by means of a dippy rhyme. No wonder Julius Caesar saw need to reform the calendar, and then Pope Gregory XIII saw need to reform Caesar's reformation, chopping eleven minutes and fourteen seconds out of the year. No wonder Russia's October Revolution turns out to have happened in November. Time, as we choose to construe it, is not one of nature's immanent truths. It's a conceptual structure of pure human contrivance, like WordPerfect or mah-jongg.

What is a second? We have arbitrarily decreed it to be one sixtieth of a minute. A minute is equally arbitrary. These things don't exist in the real world of oak trees and geese and the sun and the moon. An hour? One-twelfth of an equatorial day and one-twelfth of a night, likewise for no particular reason. Twelve seems to have been an appealing figure that, unlike ten, contains even multiples of the magical number three. In medieval European monasteries, where some standard of promptness was important for scheduled prayers, daylight was divided (before the appearance of those first mechanical clocks) at all times of year into twelve equal segments, known as the canonical hours. As a result, the twelve daytime hours were each shorter in winter than in summer. The nights in winter were longer, and therefore the night hours were longer too. Such god-like tinkering was only the logical antecedent to daylight savings, another signal of time's relativity. And of course time zones represent a small earthly pantomime of the reality of Einstein's space-time, whereby the question *when?* still can't be detached from the question *how far away?*, its Siamese twin. On an intergalactic scale, that linkage shows up in talk about light-year distances. How far to a certain star or galaxy? The answer is given as so many years' worth of flying photons. Four light-years to Alpha Centauri. Nine light-years to Sirius. Eight hundred thousand light-years, a far piece, to the Andromeda Nebula. Distance becomes time, at the scale on which relativity is noticeable, and time becomes the trajectory of light.

The philosopher Krishnamurti once offered an epigram that seems to converge here with relativity theory. "To meditate is to transcend time," he wrote. "Time is the distance that thought travels in its achievements."

What is a decade? At the speed of light, it comes to fifty-eight trillion miles of distance. But to a pious Druid who worshipped at the clock called Stonehenge, a decade would have been ten fingers' worth of spring equinoxes, each one greeted with glad clamor when the sun rose at just the right notch in the rocks. To a Crow or a Blackfeet in the northern Rockies, a decade was ten fingers' worth of mnemonic winter counts: 1804 *Heavy spring snow, even the dogs went snow-blind;* 1847 *White buffalo, Dusk killed it;* 1872 *Comes Out of the Water, she ran off the Assiniboine horses;* 1973 *Geese, they flew over in a storm.* And what about fifty years? One handful (five fingers' worth) of decades. Within the pectoral fin of a whale lurk five metacarpaloid bones that would seem to salute the fundamental rightness of a system of fives and tens; but, then again, to an octopus fifty means nothing. On dry land, we humans are left to make of it what we can.

STEPHEN HAWKING, who may be the most original thinker in physics since Einstein, is entitled by his record of brilliance to ask a question that would land most of us in a rubber room: "Why do we remember the past but not the future?"

This is a matter more puzzling than it might seem, because the laws of science, as Hawking notes, "do not distinguish between the forward and backward directions of time." The descriptions of events and transformations offered by physics, in particular, contain no inherent distinction between past and future. "Yet there is a big difference between the forward and backward directions of real time in ordinary life," he admits in his book *A Brief History of Time.* The solution that Hawking offers for this conundrum is more elaborate than satisfying, I warn you. It's based on what Hawking calls "three different arrows of time."

These are the thermodynamic arrow, the psychological arrow, and the cosmological arrow. The cosmological arrow involves arguments about how an expanding universe must differ from a collapsing universe, and if we followed him out to the turnaround point on that line of thought we would probably never get home. The thermodynamic and the psychological arrows are more pertinent here. The thermodynamic arrow of time, derived from the second law of thermodynamics, is defined by Hawking as "the direction of time in which disorder or entropy increases." In the universe that we happen to inhabit, a wineglass might fall off a table and shatter to pieces on the floor; but pieces of glass don't fall off the floor and tumble upward into conjoinment as a wineglass on a table. Therefore the wineglass is necessarily past and the fragments are necessarily future. The psychological arrow of time is "the direction in which we feel time passes, the direction in which we remember the past but not the future." That psychological arrow, Hawking says, is determined within our brains by the thermodynamic arrow. Like the memory function in a computer, creating small pockets of order within a larger context of expended electricity and dissipated heat, we too are programmed to remember things in the order in which entropy increases. Of course, common sense hollers out that we're programmed to remember things in the order in which they *happened.* But why did they happen in that order? Besides, Hawking's business is deep physics, not common sense.

Show someone a film of a falling and shattering wineglass. Then show the same film in reverse. The person will be able to tell you which showing corresponds to the flow of time. This linkage between the psychological arrow of time and the thermodynamic arrow, if I read Hawking correctly, results from the expectations that every human inherits and acquires as part of life in an entropic (and expanding) universe.

What do we measure when we measure time? The gloomy answer from Hawking, one of our most implacably cheerful scientists, is that we measure entropy. We measure changes and those changes are all for the worse. We measure increasing disorder. Life is hard, says science, and constancy is the greatest of miracles.

THE JOKE at the banquet is this: There are *two* clocks. The Irish bride and the Norwegian groom have already bought one for themselves.

The son arrived first at the midwestern airport, after weeks of long-distance consultation with his sisters, during which the banquet had been arranged and the Kern & Söhne gift clock had been settled upon. Before they were out of the parking garage his parents had confessed gleefully that they'd decided to treat themselves to a present. Something that we could choose together and share, they explained. Not a jade necklace then? he guessed. No. Not a new car? No. A clock! said his red-haired mother. A beautiful gold-and-brass clock, with lovely chimes. That's a very apt choice, said the straightfaced son. Somehow it seems just right. I can't tell you how apt it seems, he said. Later he took his own consolatory enjoyment in watching each of his sisters receive the same glad news with the same sort of maniacal, paralyzed, grinning wince.

Everyone admired the new clock. Each time its chimes sounded, the Irish bride's face lit with pleasure. The Norwegian groom curled up a corner of his mouth. They were still sharing the music of time, as they had for exactly five decades. Then everyone dressed up nice and went off to the banquet.

The banquet is not past or future; the banquet is now.

Photographs are taken. Food is eaten. Several short gentle speeches are made, and a son-in-law reads a warm comical poem. A few favorite anecdotes are brought out and polished again, for the zillionth time, like the family silver. Thanks are offered skyward. Finally the daughters and son set out the big gift-wrapped box. The joke that they've played on themselves can now be extended to their parents.

The small red-haired woman unwraps it. She lifts out the second clock, the one without chimes, the one with the twirling crystal pendulum. Einstein himself said that two clocks are better than one. She hoots. Then she tilts her head back and laughs, bravely, unregrettingly, joyfully, in the face of time.

# *Strawberries Under Ice*

## *1. The Gradient of Net Mass Balance*

Antarctica is a gently domed continent squashed flat, like a dent in the roof of a Chevy, by the weight of its ice. That burden of ice amounts to seven million cubic miles. Melt it away and the Antarctic interior would bounce upward; Earth itself would change shape. This grand cold fact has, to me, on the tiny and personal scale, a warm appeal. Take away ice and the topography of my own life changes drastically too.

Ice is lighter than water but still heavy. The stuff answers gravity. Ice is a solid but not an absolute solid. The stuff flows. Slowly but inexorably it runs downhill. We think of iciness as a synonym for cold, but cold is relative and ice happens to function well as insulation against heat loss: low thermal conductivity. Also it *releases* heat to immediate surroundings in the final stage of becoming frozen itself. Ice warms. On a certain night, roughly thirteen years ago, it warmed me.

When a tongue of ice flows down a mountain valley, we call it a glacier. When it flows out in all directions from a source point at high elevation, like pancake batter poured on a griddle, we call it a sheet. Out at the Antarctic circumference are glaciers and seaborne shelves, from which icebergs calve off under their own weight. Both sheets and glaciers are supplied with their substance, their impetus, their ice, by snow and other forms of precipitation back uphill at the source. While old ice is continually lost by calving and melting in the lowlands, new ice is deposited in the highlands, and any glacier or sheet receiving more new ice than it loses old, through the course of a year, is a glacier or sheet that is growing. The scientists would say that its net mass balance is positive.

The Antarctic sheet, for instance, has a positive balance. But this is not an essay about Antarctica.

Each point on a great ice body has its own numerical value for mass balance. Is the ice right here thicker or thinner than last year? Is the glacier, at this spot, thriving or dying? The collective profile of all those individual soundings—more ice or less? thriving or dying?—is called the gradient of net mass balance. This gradient tells, in broad perspective, what has been lost and what has been gained. On that certain night, thirteen years ago, I happened to be asking myself exactly the same question: *What's been lost and what, if anything, gained?* Because snow gathers most heavily in frigid sky-scraping highlands, the gradient of net mass balance correlates steeply with altitude. Robust glaciers come snaking down out of the Alaskan mountains. Also because snow gathers most heavily in frigid sky-scraping highlands, I had taken myself on that day to a drifted-over pass in the Bitterroot Mountains, all hell-and-gone up on the state border just west of the town of Tarkio, Montana, and started skiing uphill from there.

I needed as much snow as possible. I carried food and a goose-down sleeping bag and a small shovel. The night in question was December 31, 1975.

I hadn't come to measure depths or to calculate gradients. I had come to insert myself into a cold white hole. First, of course, I had to dig it. This elaborately uncomfortable enterprise seems to have been part of a long foggy process of escape and purgation, much of which you can be spared. Suffice to say that my snow cave, to be dug on New Year's Eve into a ten-foot-high cornice on the leeward side of the highest ridge I could ski to, and barely large enough for one person, would be at the aphelion of that long foggy process. At the perihelion was Oxford University.

At Oxford University during one week in late springtime there is a festival of crew races on the river, featuring girls in long dresses, boys in straw hats, an abundance of champagne and strawberries. This event is called Eights Week, for the fact of eight men to a crew. It is innocent. More precisely: It is no more obnoxious, no more steeped in snobbery and dandified xenophobia and intellectual and social complacence than any other aspect of Oxford University. The strawberries are served under heavy cream. Sybaritism is mandatory. For these and other reasons, partly personal, partly political, I had fled the place screaming during Eights Week of 1972, almost exactly coincident (by no coincidence) with Richard Nixon's announcement of the blockade of Haiphong harbor. Nixon's blockade and Oxford's strawberries had nothing logically in common, but they converged to produce in me a drastic reaction to what until then had been just a festering distemper.

It took me another year to arrive in Montana. I had never before set foot in the state. I knew no one there. But I had heard that it was a place where, in the early weeks of September, a person could look up to a looming horizon and see fresh-fallen snow. I had noted certain blue lines on a highway map, knew the lines to be rivers, and imagined those rivers to be dark mountain streams flashing with trout. I arrived during the early weeks of September and lo it was all true.

I took a room in an old-fashioned boardinghouse. I looked for a job. I started work on a recklessly ambitious and doomed novel. I sensed rather soon that I hadn't come to this place temporarily. I began reading the writers—Herodotus, Euripides, Coleridge, Descartes, Rousseau, Thoreau, Raymond Chandler—for whom a conscientious and narrow academic education had left no time. I spent my nest egg and then sold my Volkswagen bus for another. I learned the clownish mortification of addressing strangers with: "Hi, my name is Invisible and I'll be your waiter tonight." I was twenty-six, just old enough to realize that this period was not some sort of prelude to my life but the thing itself. I knew I was spending real currency, hard and finite, on a speculative venture at an unknowable rate of return: the currency of time, energy, stamina. Two more years passed before I arrived, sweaty and chilled, at that high cold cornice in the Bitterroots.

By then I had made a small handful of precious friends in this new place, and a menagerie of acquaintances, and I had learned also to say: "You want that on the rocks or up?" Time was still plentiful but stamina was low. Around Christmas that year, two of the precious friends announced a New Year's Eve party. Tempting, yet it seemed somehow a better idea to spend the occasion alone in a snow cave.

So here I was. There had been no trail up the face of the ridge and lifting my legs through the heavy snow had drenched and exhausted me. My thighs felt as though the Chicago police had worked on them with truncheons. I dug my hole. That done, I changed out of the soaked, freezing clothes. I boiled and ate some noodles, drank some cocoa; if I had been smart enough to encumber my pack with a bottle of wine, I don't remember it. When dark came I felt the nervous exhilaration of utter solitude and, behind that like a metallic aftertaste, loneliness. I gnawed on my thoughts for an hour or two, then retired. The night turned into a clear one and when I crawled out of the cave at 3:00 A.M. of the new year, to empty my bladder, I found the sky rolled out in a stunning pageant of scope and dispassion and cold grace.

It was too good to waste. I went back into the cave for my glasses.

The temperature by now had gone into the teens below zero. I stood there beside the cornice in cotton sweatpants, gaping up. "We never know what we have lost, or what we have found," says America's wisest poet, Robert Penn Warren, in the context of a meditation about John James Audubon and the transforming power of landscape. We never know what we have lost, or what we have found. All I did know was that the highway maps called it Montana, and that I was here, and that in the course of a life a person could travel widely but could truly open his veins and his soul to just a limited number of places.

After half an hour I crawled back into the cave, where ten feet of snow and a rime of ice would keep me warm.

## 2. *Ablation*

Trace any glacier or ice sheet downhill from its source and eventually you will come to a boundary where the mass balance of ice is zero. Nothing is lost, over the course of time, and nothing is gained. The ice itself constantly flows past this boundary, molecule by molecule, but if any new ice is added here by precipitation, if any old ice is taken away by melting, those additions and subtractions cancel each other exactly. This boundary is called the equilibrium line. Like other forms of equilibrium, it entails a cold imperturbability, a sublime steadiness relative to what's going on all around. Above the equilibrium line is the zone of accumulation. Below is the zone of ablation.

*Ablation* is the scientists' fancy word for loss. Down here the mass balance is negative. Ice is supplied to this zone mainly by flow from above, little or not at all by direct precipitation, and whatever does come as direct precipitation is less than the amount annually lost. The loss results from several different processes: wind erosion, surface melting, evaporation (ice does evaporate), underside melting of an ice shelf where it rests on the warmer sea water. Calving off of icebergs. *Calving* is the scientists' quaint word for that sort of event when a great hunk of ice—as big as a house or, in some cases, as big as a county—tears away from the leading edge of the sheet or the glacier and falls thunderously into the sea.

Possibly this talk about calving reflects an unspoken sense that the larger ice mass, moving, pulsing, constantly changing its shape, is almost alive. If so, the analogy doesn't go far. Icebergs don't suckle or grow. They float away on the sea, melt, break apart, disappear. Wind erosion and evaporation and most of those other ablative processes work on the ice slowly, incrementally. Calving on the other hand is abrupt. A large piece of the whole is there, and then gone.

The occurrence of a calving event depends on a number of factors—flow rate of the whole ice body, thickness at the edge, temperature, fissures in the ice, stresses from gravity or tides—one of which is the strength of the ice itself. That factor, strength, is hard to measure. You might never know until too late. Certain experiments done on strength-testing machines have yielded certain numbers: a strength of thirty-eight bars (a bar is a unit of pressure equal to 100,000 newtons per square meter) for crushing; fourteen bars for bending; nine bars for tensile. But those numbers offer no absolute guide to the performance of different types of ice under different conditions. They only suggest in a relative way that, though ice may flow majestically under its own weight, though it may stretch like caramel, though it may bend like lead, it gives back rock-like resistance to a force coming down on it suddenly. None of this cold information was available to me on the day now in mind, and if it had been I wouldn't have wanted it.

On the day now in mind I had been off skiing, again, with no thought for the physical properties of ice, other than maybe some vague awareness of the knee strain involved in carving a turn across boilerplate. I came home to find a note in my door.

The note said that a young woman I knew, the great love of a friend of mine, was dead. The note didn't say what had happened. I should call a number in Helena for details. It was not only shocking but ominous. Because I knew that the young woman had lately been working through some uneasy and confusing times, I thought of all the various grim possibilities involving despair. Then I called the Helena number, where a houseful of friends were gathered for communal grieving and food and loud music. I learned that the young woman had died from a fall. A freak accident. In the coldest sense of cold consolation, there was in this information some relief.

She had slipped on a patch of sidewalk ice, the night before, and hit her head. A nasty blow, a moment or two of unconsciousness, but she had apparently been all right. She went home alone and was not all right and died before morning. I suppose she was about twenty-seven. This is exactly why head-trauma cases are normally put under close overnight observation, but I wasn't aware of that at the time, and neither evidently were the folks who had helped her up off that icy sidewalk. She had seemed okay. Even after the fall, her death was preventable. Of course most of us will die preventable deaths; hers was only more vividly so, and earlier.

I had known her, not well, through her sweetheart and the network of friends now assembled in that house in Helena. These friends of hers and mine were mostly a group of ecologists who had worked together, during graduate school, as waiters and bartenders and cooks; I met them in that context and they had nurtured my sanity to no small degree when that context began straining it. They read books, they talked about ideas, they knew a spruce from a hemlock, they slept in snow caves: a balm of good company to me. They made the state of Montana into a place that was not only cold, true, hard, and beautiful, but damn near humanly habitable. The young woman, now dead, was not herself a scientist, but she was one of them in all other senses. She came from a town up on the Hi-Line.

I had worked with her too, and seen her enliven long afternoons that could otherwise be just a tedious and not very lucrative form of self-demeanment. She was one of those rowdy, robust people—robust in good times, just as robust when she was angry or miserable—who are especially hard to imagine dead. She was a rascal of wit. She could be wonderfully crude. We all knew her by her last name, because her first seemed too ladylike and demure. After the phone call to Helena, it took me a long time to make the mental adjustment of tenses. She *had* been a rascal of wit.

The memorial service was scheduled for such-and-such day, in that town up on the Hi-Line.

We drove up together on winter roads, myself and two of the Helena friends, a husband-and-wife pair of plant ecologists. Others had gone ahead. Places available for sleeping, spare rooms and floors; make contact by phone; meet at the church. We met at the church and sat lumpish while a local pastor discoursed with transcendent irrelevance about what we could hardly recognize as her life and death. It wasn't his fault, he didn't know her. There was a reception with the family, followed by a postwake on our own at a local bar, a fervent gathering of young survivors determined not only to cling to her memory but to cling to one another more appreciatively now that such a persuasive warning bell of mortality had been rung, and then sometime after dark as the wind came up and the temperature dropped away as though nothing was under it and a new storm raked in across those wheatlands, the three of us started driving back south. It had been my first trip to the Hi-Line.

Aside from the note in the door, this is the part I remember most clearly. The car's defroster wasn't working. I had about four inches of open windshield. It was a little Honda that responded to wind like a shuttlecock, and on slick pavement the rear end flapped like the tail of a trout. We seemed to be rolling down a long dark tube coated inside with ice, jarred back and forth by the crosswinds, nothing else visible except the short tongue of road ahead and the streaming snow and the trucks blasting by too close in the other lane. How ironic, I thought, if we die on the highway while returning from a funeral. I hunched over the wheel, squinting out through that gap of windshield, until some of the muscles in my right shoulder and neck shortened themselves into a knot. The two plant ecologists kept me awake with talk. One of them, from the backseat, worked at the knot in my neck. We talked about friendship and the message of death as we all three felt we had heard it, which was to cherish the living, while you have them. Seize, hold, appreciate. Pure friendship, uncomplicated by romance or blood, is one of the most nurturing human relationships and one of the most easily taken for granted. This was our consensus, spoken and unspoken.

These two plant ecologists had been my dear friends for a few years, but we were never closer than during that drive. Well after midnight, we reached their house in Helena. I slept on sofa cushions. In the morning they got me to a doctor for the paralytic clench in my neck. That was almost ten years ago and I've hardly seen them since.

The fault is mine, or the fault is nobody's. We got older and busier and trails diverged. They began raising children. I traveled to Helena less and less. Mortgages, serious jobs, deadlines; and the age of sleeping on sofa cushions seemed to have passed. I moved, they moved, opening more geographical distance. Montana is a big place and the roads are often bad. These facts offered in explanation sound even to me like excuses. The ashes of the young woman who slipped on the ice have long since been sprinkled onto a mountaintop or into a river, I'm not sure which. Nothing to be done now either for her or about her. The two plant ecologists I still cherish, in intention anyway, at a regrettable distance, as I do a small handful of other precious friends who seem to have disappeared from my life by wind erosion or melting.

### 3. *Leontiev's Axiom*

The ice mass of a mountain glacier flows down its valley in much the same complicated pattern as a river flowing in its bed. Obviously the glacier is much slower. Glacial ice may move at rates between six inches and six feet per day; river water may move a distance in that range every second. Like the water of a river, though, the ice of a glacier does not all flow at the same rate. There are eddies and tongues and slack zones, currents and swells, differential vectors of mix and surge. The details of the flow pattern depend on variable parameters special to each case: depth of the ice, slope, contour of the bed, temperature. But some generalizations can be made. Like a river, a glacier will tend to register faster flow rates at the surface than at depths, faster flows at midchannel than along the edges, and faster flows down in the middle reaches than up near the source. One formula scientists use to describe the relations between flow rate and those other factors is:

$$u = k_1 \sin^3 a\, h^4 + k_2 \sin^2 a\, h^2.$$

Everyone stay calm. This formula is not Leontiev's Axiom, and so we aren't going to bother deciphering it.

Turbulent flow is what makes a glacier unfathomable, in the sense of *fathoming* that connotes more than taking an ice-core measurement of depth. Turbulent flow is also what distinguishes a river from, say, a lake. When a river freezes, the complexities of turbulent flow interact with the peculiar physics of ice formation to produce a whole rat's nest of intriguing and sometimes inconvenient surprises. Because of turbulence, the water of a river cools down toward the freezing point uniformly, not in stratified layers as in a lake. Eventually the entire mass of flowing water drops below thirty-two degrees Fahrenheit. Small disks of ice, called frazil ice, then appear. Again because of turbulence, this frazil ice doesn't all float on the surface (despite being lighter than water) but mixes throughout the river's depth. Frazil ice has a tendency toward adhesion, so some of it sticks to riverbed rocks. Some of it gloms onto bridge pilings and culverts, growing thick as a soft cold fur. Some of it aggregates with other frazil ice, forming large dollops of drifting slush. Meanwhile, huge slabs of harder sheet ice, formed along the banks and broken free as the river changed level, may also be floating downstream. The slabs of sheet ice and the dollops of frazil ice go together like bricks and mortar. Stacking up at a channel constriction, they can lock themselves into an ice bridge.

Generally, when such an ice bridge forms, the river will have room to flow underneath. But if the river is very shallow and the slabs of sheet ice are large, possibly not. Short of total blockage, the flow of the river will be slowed where it must pass through that narrowed gap; if it slows to less than some critical value, more ice will collect along the front face of the bridge and the ice cover will expand upstream. The relevant formula here is:

$$V_c = (1 - h/H)\sqrt{2g(p - p_i/p)h}$$

where $V_c$ is the critical flow rate and $h$ is the ice thickness and everything else represents something too. But this also is not Leontiev's Axiom, and so we can ignore it, praise God.

The Madison River where it runs north through Montana happens to be very shallow. Upstream from (that is, south of) the lake that sits five miles north of the small town of Ennis, it's a magnificent stretch of habitat for stoneflies and caddisflies and trout and blue heron and fox and eagles and, half the year anyway, fishermen. The water is warmed at its geothermal source in Yellowstone Park, cooled again by its Montana tributaries such as West Fork, rich in nutrients and oxygen, clear, lambent, unspoiled. Thanks to these graces, it's probably much too famous for its own good, and here I am making it a little more famous still. Upstream from the highway bridge at Ennis, where it can be conveniently floated by fishermen in rafts and guided Mackenzie boats, it gets an untoward amount of attention. This is where the notorious salmonfly hatch happens: boat traffic like the Henley Regatta, during that dizzy two weeks of June while the insects swarm and the fish gluttonize. This is the stretch of the Madison for fishermen who crave trophies but not solitude. Downstream from the Ennis bridge it becomes a different sort of river. It becomes a different sort of place.

Downstream from the Ennis bridge, for that five-mile stretch to the lake, the Madison is a broken-up travesty of a river that offers mediocre fishing and clumsy floating and no trophy trout and not many salmonflies and I promise you fervently you wouldn't like it. This stretch is called the channels. The river braids out into a maze of elbows and sloughs and streams separated by dozens of small and large islands, some covered only with grass and willow, some shaded with buckling old cottonwoods, some holding thickets of water birch and woods rose and raspberry scarcely tramped through by a single fisherman in the course of a summer. The deer love these islands and, in May, so do the nesting geese. Mosquitoes are bad here. The walking is difficult and there are bleached cottonwood deadfalls waiting to tear your waders. At the end of a long day's float, headwinds and choppy waves will come up on the lake just as you try to row your boat across to the ramp. Take my word, you'd hate the whole experience. Don't bother. Give it a miss. I adore that five miles of river more than any other piece of landscape in the world.

Surrounding the braidwork of channels is a zone of bottomland roughly two miles wide, a great flat swatch of subirrigated meadow only barely above the river's springtime high-water level. This low meadow area is an unusual sort of no-man's-land that performs a miraculous service: protecting the immediate riparian vicinity of the channels from the otherwise-inevitable arrival of ranch houses, summer homes, resort lodges, motels, paved roads, development, spoliation, and all other manner of venal doom. Tantalizing and vulnerable as it may appear on a July afternoon, the channels meadowland is an ideal place to raise bluegrass and Herefords and sandhill cranes but, for reasons we'll come to, is not really good for much else.

By late December the out-of-state fishermen are long gone, the duck hunters more recently, and during a good serious stretch of weather the dark river begins to flow gray and woolly with frazil ice. If the big slabs of sheet ice are moving too, a person can stand on the Ennis highway bridge and hear the two kinds of ice rubbing, hissing, whispering to each other as though in conspiracy toward mischief—or maybe revenge. (Through the three years I lived in Ennis myself, I stood on that bridge often, gawking and listening. There aren't too many other forms of legal amusement in a Montana town of a thousand souls during the short days and long weeks of midwinter.) By this time the lake, five miles downstream, will have already frozen over. Then the river water cools still further, the frazil thickens, the slabs bump and tumble into those narrow channels, until somewhere, at a point of constriction down near the lake, mortar meets brick and you begin to get:

$$V_c = (1 - h/H)\sqrt{2g(p - p_i/p)h}.$$

Soon the river is choked with its own ice. All the channels are nearly or totally blocked. But water is still arriving from upstream, and it has to go somewhere. So it flows out across the bottomland. It spills over its banks and, moving quickly, faster than a man can walk, it covers a large part of that meadow area with water. Almost as quickly, the standing floodwater becomes ice.

If you have been stubborn or foolish enough to build your house on that flat, in a pretty spot at the edge of the river, you now have three feet of well-deserved ice in your living room. *Get back away from me,* is what the river has told you. *Show some goddamn respect.* There are memories of this sort of ice-against-man encounter. It hasn't happened often, that a person should come along so mule-minded as to insist on flouting the reality of the ice, but often enough for a few vivid stories. Back in 1863, for instance, a settler named Andrew Odell, who had built his cabin out on the channel meadows, woke up one night in December to find river water already lapping onto his bed. He grabbed his blanket and fled, knee deep, toward higher ground on the far side of a spring creek that runs parallel to the channels a half mile east. That spring creek is now called Odell Creek, and it marks a rough eastern boundary of the zone that gets buried in ice. Nowadays you don't see any cabins or barns in the flat between Odell Creek and the river.

Folks in Ennis call this salutary ice-laying event the Gorge. The Gorge doesn't occur every year, and it isn't uniform or predictable when it does. Two or three winters may go by without serious weather, without a Gorge, without that frozen flood laid down upon hundreds of acres, and then there will come a record year. A rancher named Ralph Paugh remembers one particular Gorge, because it back-flooded all the way up across Odell Creek to fill his barn with a two-foot depth of ice. This was on Christmas Day, 1983. "It come about four o'clock," he recalls. "Never had got to the barn before." His barn has sat on that rise since 1905. He has some snapshots from the 1983 episode, showing vistas and mounds of whiteness. "That pile there, see, we piled that up with the dozer when we cleaned it out." Ralph also remembers talk about the Gorge in 1907, the year he was born. That one took out the old highway bridge, so for the rest of the winter schoolchildren and mailmen and whoever else had urgent reason for crossing the river did so on a trail of planks laid across ice. The present bridge is a new one, the lake north of Ennis is also a relatively recent contrivance (put there for hydroelectric generation about the time, again, when Ralph Paugh was born), but the Gorge of the Madison channels is natural and immemorial.

I used to lace up my Sorels and walk on it. Cold sunny afternoons of January or February, bare willows, bare cottonwoods, exquisite solitude, fox tracks in an inch of fresh snow, and down through three feet of ice below my steps and the fox tracks were spectacular bits of Montana that other folk, outlanders, coveted only in summer.

Mostly I wandered these places alone. Then one year a certain biologist of my recent acquaintance came down for a visit in Ennis. I think this was in late April. I know that the river had gorged that year and that the ice was now melting away from the bottomland, leaving behind its moraine of fertile silt. The channels themselves, by now, were open and running clear. The first geese had arrived. This biologist and I spent that day in the water, walking downriver through the channels. We didn't fish. We didn't collect aquatic insects or study the nesting of *Branta canadensis.* The trees hadn't yet come into leaf and it was no day for a picnic. We just walked in the water, stumbling over boulders, bruising our feet, getting wet over the tops of our waders. We saw the Madison channels, fresh from cold storage, before anyone else did that year. We covered only about three river miles in the course of the afternoon, but that was enough to exhaust us, and then we stumbled out across the muddy fields and walked home on the road. How extraordinary, I thought, to come across a biologist who would share my own demented appreciation of such an arduous, stupid, soggy trek. So I married her.

The channels of the Madison are a synecdoche. They are the part that resonates suggestively with the significance of the whole. To understand how I mean that, it might help to know Leontiev's Axiom. Konstantin Leontiev was a cranky nineteenth-century Russian thinker. He trained as a physician, worked as a diplomat in the Balkans, wrote novels and essays that aren't read much today, became disgusted at the prospect of moral decay in his homeland, and in his last years flirted with becoming a monk. By most of the standards you and I likely share, he was an unsavory character. But even a distempered and retrograde Czarist with monastic leanings can be right about something once in a while.

Leontiev wrote: "To stop Russia from rotting, one would have to put it under ice."

In my mind, in my dreams, that great flat sheet of Madison River whiteness spreads out upon the whole state of Montana. I believe, with Leontiev, in salvation by ice.

### 4. *Sources*

The biologist whose husband I am sometimes says to me: "All right, so where do we go when Montana's been ruined? Alaska? Norway? Where?" This is a dark joke between us. She grew up in Montana, loves the place the way some women might love an incorrigibly self-destructive man, with pain and fear and pity, and she has no desire to go anywhere else. I grew up in Ohio, discovered home in Montana only fifteen years ago, and I feel the same. But still we play at the dark joke. "Not Norway," I say, "and you know why." We're each half Norwegian and we've actually eaten lutefisk. "How about Antarctica," I say. "Antarctica should be okay for a while yet."

On the desk before me now is a pair of books about Antarctica. Also here are a book on the Arctic, another book titled *The World of Ice*, a book of excerpts from Leontiev, a master's thesis on the subject of goose reproduction and water levels in the Madison channels, an extract from an unpublished fifty-year-old manuscript on the history of the town of Ennis, a cassette tape of a conversation with Ralph Paugh, and a fistful of photocopies of technical and not-so-technical articles. One of the less technical articles is titled "Ice on the World," from a recent issue of *National Geographic.* In this article is a full-page photograph of strawberry plants covered with a thick layer of ice.

These strawberry plants grew in central Florida. They were sprayed with water, says the caption, because subfreezing temperatures had been forecast. The growers knew that a layer of ice, giving insulation, even giving up some heat as the water froze, would save them.

In the foreground is one large strawberry. The photocopy shows it dark gray, but in my memory it's a death-defying red.

# *Notes and Provenance*

Although each of the essays in this collection has previously appeared in a magazine or anthology, and although each was in some degree shaped by the timing and circumstances of its first publication, I've tried to include only pieces that don't feel stale. None of them has been entirely updated into the present, in terms of the arithmetic of elapsed years and personal milestones, but most of their immediate chronological references have been deemphasized. Still, in some cases the original context or occasion—for instance, Edward Abbey's death in 1989, which elicited the eulogy titled "Bagpipes for Ed"—bears recollecting. Also, in several of these pieces I've taken an alternate look at subjects, persons, or places mentioned in my 1996 book, *The Song of the Dodo.* So I'll note here some contextual particulars regarding a few of the essays and then provide a basic first-publication record for all of them.

"SYNECDOCHE and the Trout" is placed at the start of this collection partly because it's the oldest piece included. My admiration for trout as animals remains unabated, but my fly rods now gather dust in the basement. The scientific names *Oncorhynchus clarki, Oncorhynchus mykiss,* and *Salmo trutta* refer respectively to cutthroat trout, rainbow trout, and brown trout.

The first two of those binomials, *O. clarki* and *O. mykiss*, are recent revisions, having been officially adopted by scientific adjudicators in 1989, after Gerald R. Smith and Ralph F. Stearley published a paper (see bibliography) demonstrating that cutthroat and rainbow trout are actually more closely related to Pacific salmon (genus *Oncorhynchus*) than to brown trout and Atlantic salmon (genus *Salmo*). Previously the cutthroat and the rainbow had been called *Salmo clarki* and *Salmo gairdneri,* which is how I knew them when I was a fishing guide. It should also be noted that neither brown trout nor rainbows are native to Montana rivers in the biogeographical sense; then again, I'm not native to Montana either, so who am I to carp? Another change (gustatorily welcome but culturally ominous) since this piece was first published, back in 1988: Espresso *is* now available in Montana. The "first trout I ever caught," as mentioned herein, is not to be confused with the fish that fed me the night I was stuck out near Fool's Creek, as described in the introduction; roughly a week of fevered fishing had intervened.

"The White Tigers of Cincinnati" received a Genesis Award from The Ark Trust, Inc., of Studio City, California, in 1991.

"Superdove on 46th Street" was my final "Natural Acts" column in *Outside*, concluding a sequence of roughly 160 essays I'd published under that rubric between 1981 and 1996. In the original version, the penultimate paragraph included a passing announcement of my retirement and a brief salutation to the "good and valued readers" with whom I'd shared such a long-term relationship; the statement about "leaving you with something to chew on" took its first meaning from those circumstances.

"The Keys to Kingdom Come," as mentioned in the footnote, was written in 1987, when the global balance of thermonuclear terror was more dualistic, and probably more precarious, than now. I still have on my office wall a Montana map from that time, showing the state pocked with scores of red dots, each dot representing an ICBM silo. I applied the red dots myself, based on unclassified but official miliary information, and I hung the map as a reminder against insouciance in this seemingly remote, seemingly halcyon state. Insouciance is ill-advised, any place, any time. Although things have changed somewhat in terms of the number and targeting of American ICBMs, and although I haven't done any research on the particulars of those changes, my assumption is that young men and women from Malmstrom Air Force Base still pull alerts underground, armed with keys capable of launching nuclear missiles, if not toward the USSR then somewhere else.

"The Trees Cry Out on Currawong Moor" incorporates some historical material that was also treated, in a different form and context, within *The Song of the Dodo.*

"Eat of This Flesh" is the second of two essays I've written on the subject of mountain lions. The first, titled "Five Kinds of Rarity" when it appeared in *Outside* in May 1990, was reprinted as part of a special tabloid on biological diversity published by the Montana Wilderness Association in November 1992. That's where it caught the attention, and provoked the ire, of Don Thomas. I haven't chosen to reprint that earlier essay in this book because (1) as Thomas pointed out, it wasn't as carefully considered or as grounded in knowledge as it should have been, and (2) "Eat of This Flesh" has preempted it.

"Trinket from Aru" includes a set of natural-history observations that I built up to, but purposefully refrained from presenting, in *The Song of the Dodo:* what I saw on a certain ridge-top in the Aru Islands where the greater bird of paradise, *Paradisaea apoda,* performs its display.

"Bagpipes for Ed" was written soon after Edward Abbey's death, in March 1989, and appeared in June of that year.

"Point of Attachment" is included within section IV because, in my view, Darwin and his barnacles offer an interesting counterpoint to Gilbert White and *Hirundo rustica,* as described in "The Swallow That Hibernates Underwater."

"Love in the Age of Relativity" was written in February 1991, and owed its existence (as I do) to the fact that Mary L. Egan and Willard A. Quammen were married, in LaCrosse, Wisconsin, on February 22, 1941. The winter counts are borrowed, gratefully, from my colleague and friend Barry Lopez.

"Strawberries Under Ice" was written in 1988, and the reference to "a certain night, roughly thirteen years ago," namely December 31, 1975, reflects the essay's original chronological perspective. Likewise, the statement that I had "discovered home in Montana only fifteen years ago" was arithmetically accurate as of 1988; in nonarithmetical terms, it remains as accurate today as ever. I arrived in the state, for the first time in my life and (though I didn't know it) to stay, on September 12, 1973. Hope abides; Norway and Antarctica are still only distant, desperate options.

The Honda described in this essay, by the way, carrying the three of us back from that Hi-Line funeral, is the same vehicle mentioned in Robert Penn Warren's poem *Chief Joseph of the Nez Perce,* for its role in carrying Warren, myself, and Stuart Wright to the Bear Paw battlefield. It lived a full life, that little yellow Civic, and covered many Montana miles.

Ralph Paugh died on February 1, 1996. He is much missed.

FINALLY, a note of acknowledgment. I've had an extraordinarily satisfying and very fruitful relationship with *Outside* magazine over the past seventeen years, and that relationship has accounted for most of the essays in this book. Many of them simply would never have been written if I'd had to pitch the ideas, one by one, to editors less imaginative and trusting than those I've known at *Outside.* So I'm deeply grateful to Mark Bryant, John

Rasmus, Greg Cliburn, Matthew Childs, and all the other (present or former) *Outside* folk who have given me creative latitude, rigorous line-editing, and various other sorts of crucial home-office help; I'm also very grateful to Larry Burke, owner and publisher of the magazine, for his long-term support of my work.

In connection with the other pieces in this book, previously published in other magazines or anthologies, I offer warm thanks to David Seybold, Skip Brown, Rebecca Farwell, Bill Beuttler, Kathy Ely, Tamara Robbins, Philip Grossberger, Geraldine Patrick, Steve Casimiro, Rob Story, Bob Wallace, Susan Murcko, Lt. Carla Sylvester, Doug Crichton, Geoffrey Wolff, Robert Atwan, Norman Sims, Mark Kramer, Bill Kittredge, and Jan Konigsberg for their various acts of incitement, support, editing, translating, assistance, and good will. Maria Guarnaschelli and Renée Wayne Golden have my thanks for guiding this collection into existence as a book.

I'm also vastly grateful, of course, to those brave, trusting, and colorful individuals—such as Chris Spelius, Dave Foreman, Michael Soulé, Dickie Hall, John Cotter, Conrad Strickland, Karl Birkeland, Don Thomas, and others—who granted me literary access to their lives, their works, and some of their more private thoughts. Although journalism shouldn't be overly concerned with pleasing the people who happen to be spotlighted, still it's important to keep faith with sources; that involves accuracy and fairness, not flattery. I hope the people who trusted me most will have the least cause to regret.

Kris Ellingsen was the first reader, the first literary and scientific critic, of virtually all of these pieces. Her brain and sensibility have helped inform them in ways beyond describing.

First publication of each of the pieces was as follows:

"Synecdoche and the Trout" appeared as a chapter in *Seasons of the Angler*, edited by David Seybold (New York: Weidenfeld & Nicolson, 1988); "Time and Tide on the Ocoee River" (as "Spelius Rides the Wave"), *Outside* (March 1994); "Vortex," *Outside* (September 1992); "Only Connect," *Outside* (December 1993); "Grabbing the Loop" (as "Down a Crazy River"), *Destination Discovery* (December 1995); "The White Tigers of Cincinnati," *Outside* (September 1991); "To Live and Die in L.A.," *Outside* (June 1992); "Reaction Wood," *Outside* (January 1996); "Superdove on 46th Street," *Outside* (March 1996); "Before the Fall," *Outside* (November 1993); "Pinhead Secrets" (as "Cry Pinhead!"), *Outside* (December 1994); "The Keys to Kingdom Come," *Rolling Stone* (June 18, 1987); "Karl's Sense of Snow," *Outside* (December 1995); "The Trees Cry Out on Currawong Moor," *American Way* (November 1, 1990); "The Big Turn," *Powder* (December, 1995); "Eat of This Flesh," *Outside* (May 1994); "The Swallow

That Hibernates Underwater," *Outside* (July 1992); "Trinket from Aru," *Outside* (September 1993); "Bagpipes for Ed," *Outside* (June 1989); "Point of Attachment," *Outside* (October 1994); "Voice Part for a Duet," *Outside* (August 1995); "Love in the Age of Relativity," *Outside* (June 1991); "Strawberries Under Ice," *Outside* (October 1988) and as a chapter in *Montana Spaces,* edited by William Kittredge (New York: Nick Lyons Books, 1988). "Strawberries Under Ice" has also been reprinted in *The Best American Essays 1989,* edited by Geoffrey Wolff (New York: Ticknor & Fields, 1989) and in *Literary Journalism,* edited by Norman Sims and Mark Kramer (New York: Ballantine Books, 1995).

# Bibliography

This bibliography consists mainly of the sources upon which I relied when each of the various pieces was first written. In addition, though, it also includes a few relevant works published more recently, mentioned here to help guide further investigations by any fervently curious readers.

### SYNECDOCHE AND THE TROUT


Hynes, H.B.N. 1970. *The Ecology of Running Waters.* Toronto: University of Toronto Press.

Smith, Gerald R., and Ralph F. Stearley. 1989. "The Classification and Scientific Names of Rainbow and Cutthroat Trouts." *Fisheries,* Vol. 14, No. 1.

Usinger, Robert L., ed. 1956. *Aquatic Insects of California.* Berkeley: University of California Press.

Wilde, Oscar. 1973. "The Ballad of Reading Gaol." In *De Profundis and Other Writings.* Harmondsworth, Middlesex, England: Penguin Books.

## TIME AND TIDE ON THE OCOEE RIVER

What Chris Spelius does and the way he does it is generally inaccessible through literary channels. But you could look at *Performance Kayaking*, by Stephen B. U'Ren (Harrisburg, Pa.: Stackpole Books, 1990). Or contact Spelius's own company, Expediciones Chile, at (704) 488-9082 in Bryson City, North Carolina, to obtain a video or a brochure.

## VORTEX


Bellhouse, B.J., and L. Talbot. 1969. "The Fluid Mechanics of the Aortic Valve." *Journal of Fluid Mechanics*, Vol. 35, Part 4.

Bronowski, J. 1969. "Leonardo da Vinci." In *The Penguin Book of the Renaissance*. J.H. Plumb, with essays by Garrett Mattingly et al. Harmondsworth, Middlesex, England: Penguin Books.

Clark, Kenneth. 1973. *Leonardo da Vinci: An Account of His Development as an Artist.* Harmondsworth, Middlesex, England: Penguin Books.

———. 1964. *The Drawings of Leonardo da Vinci.* With an introduction and notes by A.E. Popham. London: Jonathan Cape.

Cook, Theodore Andrea. 1979. *The Curves of Life: Being an Account of Spiral Formations and Their Application to Growth in Nature, to Science and to Art.* New York: Dover Publications, Inc. Reprint of the 1914 edition.

Daily, James W., and Donald R.F. Harleman. 1966. *Fluid Dynamics.* Reading, Mass.: Addison-Wesley Publishing Company.

Dryden, Hugh L., Francis D. Murnaghan, and H. Bateman. 1956. *Hydrodynamics.* New York: Dover Publications, Inc.

Gould, Cecil. 1975. *Leonardo: The Artist and the Non-Artist.* Boston: New York Graphic Society.

Heydenreich, Ludwig Heinrich. 1951. *Leonardo da Vinci the Scientist.* International Business Machines Corporation.

Kemp, Martin. 1981. *Leonardo da Vinci: The Marvellous Works of Nature and Man.* Cambridge, Mass.: Harvard University Press.

Leonardo da Vinci. 1981. *Leonardo da Vinci Nature Studies from the Royal Library at Windsor Castle.* Catalogue by Carlo Pedretti, Introduction by Kenneth Clark. New York: Harcourt Brace Jovanovich, Publishers.

———. 1970. *The Notebooks of Leonardo da Vinci.* Compiled and edited from the original manuscripts by Jean Paul Richter. Two vols. New York: Dover Publications, Inc. Reprint of the 1883 edition.

———. 1985. *The Complete Paintings of Leonardo da Vinci.* Introduction by L.D. Ettlinger, notes and catalogue by Angela Ottino della Chiesa. New York: Penguin Books.

Lugt, Hans J. 1983. *Vortex Flow in Nature and Technology.* New York: John Wiley & Sons.

Nealy, William. 1986. *Kayak: The Animated Manual of Intermediate and Advanced Whitewater Technique.* Birmingham, Al.: Menasha Ridge Press.

———. 1981. *Whitewater Home Companion: Southeastern Rivers.* Volume I. Birmingham, AL.: Menasha Ridge Press.

Pedretti, Carlo. 1973. *Leonardo: A Study in Chronology and Style.* Berkeley: University of California Press.

Schwenk, Theodor. 1976. *Sensitive Chaos: The Creation of Flowing Forms in Water and Air.* Translated by Olive Whicher and Johanna Wrigley. New York: Schocken Books.

Stevens, Peter S. 1974. *Patterns in Nature.* Boston: Atlantic Monthly Press/Little, Brown and Company.

Truesdell, C. 1968. "The Mechanics of Leonardo da Vinci." In *Essays in the History of Mechanics.* New York: Springer-Verlag.

Vallentin, Antonina. 1952. *Leonardo da Vinci: The Tragic Pursuit of Perfection.* New York: The Viking Press.

Wallace, Robert. 1966. *The World of Leonardo.* New York: Time-Life Books.

Zammattio, Carlo. 1974. "Mechanics of Water and Stone." In *The Unknown Leonardo,* edited by Ladislao Reti. New York: McGraw-Hill Book Company.

## ONLY CONNECT

Bader, Mike. 1991. "The Northern Rockies Ecosystem Protection Act: A Citizen Plan for Wildlands Management." *Western Wildlands,* Summer 1991.

Foreman, Dave, and John Davis, eds. 1992. "The Wildlands Project." Special issue of the journal *Wild Earth.* Canton, N.Y.

Foreman, Dave, and Howie Wolke. 1989. *The Big Outside.* Tucson, Ariz.: Ned Ludd.

Forster, E.M. 1989. *Howards End.* New York: Vintage Books. Originally published in 1910.

Harris, Larry D., and Peter B. Gallagher. 1989. "New Initiatives for Wildlife Conservation: The Need for Movement Corridors." In *Preserving Communities and Corridors,* edited by Gay Mackintosh. Washington, D.C.: Defenders of Wildlife.

MacArthur, Robert H., and Edward O. Wilson. 1967. *The Theory of Island Biogeography.* Princeton, N.J.: Princeton University Press.

Mann, Charles C., and Mark L. Plummer. 1993. "The High Cost of Biodiversity." *Science,* Vol. 260, June 25, 1993.

Metzgar, Lee H., and Mike Bader. 1992. "Large Mammal Predators in the Northern Rockies: Grizzly Bears and Their Habitat." *The Northwest Environmental Journal,* Vol. 8, No. 1.

Newmark, William Dubois. 1986. "Mammalian Richness, Colonization, and Extinction in Western North American National Parks." Ph.D. dissertation, University of Michigan.

Noss, Reed F. 1987. "Corridors in Real Landscapes: A Reply to Simberloff and Cox." *Conservation Biology,* Vol. 1, No. 2.

———. 1992. "The Wildlands Project: Land Conservation Strategy." In Foreman and Davis (1992).

Quammen, David. 1996. *The Song of the Dodo.* New York: Scribner.

Simberloff, Daniel, and James Cox. 1987. "Consequences and Costs of Conservation Corridors." *Conservation Biology,* Vol. 1, No. 1.

Simberloff, Daniel, James A. Farr, James Cox, and David W. Mehlman. 1992. "Movement Corridors: Conservation Bargains or Poor Investments?" *Conservation Biology,* Vol. 6, No. 4.

Wilson, Edward O., and Edwin O. Willis. 1975. "Applied Biogeography." In *Ecology and Evolution of Communities,* edited by Martin L. Cody and Jared M. Diamond. Cambridge, Mass.: The Belknap Press of Harvard University Press.

## GRABBING THE LOOP

Bechdel, Les, and Slim Ray. 1985. *River Rescue.* Boston: Appalachian Mountain Club.

Harrison, Jim. 1973. *A Good Day to Die.* New York: Dell.

## THE WHITE TIGERS OF CINCINNATI

Bendiner, Robert. 1981. *The Fall of the Wild and the Rise of the Zoo.* New York: E.P. Dutton.

Berger, John. 1980. "Why Look at Animals?" In Berger's collection of essays, *About Looking.* New York: Pantheon Books.

Berrier, Harry H., et al. 1975. "The White Tiger Enigma." *Veterinary Medicine/Small Animal Clinician,* Vol. 70, April 1975.

Bouissac, Paul. 1976. *Circus and Culture: A Semiotic Approach.* Bloomington, Ind.: Indiana University Press.

Crandall, Lee S. 1964. *The Management of Wild Mammals in Captivity.* Chicago: The University of Chicago Press.

Guillery, R.W. 1974. "Visual Pathways in Albinos." *Scientific American,* Vol. 230, May 1974.

Guillery, R.W., and J.H. Kass. 1973. "Genetic Abnormality of the Visual Pathways in a 'White' Tiger." *Science,* Vol. 180, June 22, 1973.

Isaac, Jeannette. 1984. "Tiger Tale." *Geo,* Vol. 6, August 1984.

Latinen, Katherine. 1987. "White Tigers and Species Survival Plans." In Tilson and Seal (1987).

Leyhausen, Paul, and Theodore H. Reed. 1971. "The White Tiger: Care and Breeding of a Genetic Freak." *Smithsonian,* Vol. 2, April 1971.

Maruska, Edward J. 1987. "White Tiger: Phantom or Freak?" In Tilson and Seal (1987).

Reed, Elizabeth C. 1970. "White Tiger in My House." *National Geographic Magazine,* Vol. 137, No. 4, April 1970.

Reed, Theodore H. 1961. "Enchantress: Queen of an Indian Palace, A Rare White Tiger Comes to Washington." *National Geographic Magazine,* Vol. 119, No. 5, May 1961.

Ritvo, Harriet. 1987. *The Animal Estate: The English and Other Creatures in the Victorian Age.* Cambridge, Mass.: Harvard University Press.

Roychoudhury, A.K. 1987. "White Tigers and Their Conservation." In Tilson and Seal (1987).

Siebert, Charles. 1991. "Where Have All the Animals Gone: The Lamentable Extinction of Zoos." *Harper's,* May 1991.

Simmons, Lee G. 1987. "White Tigers: The Realities." In Tilson and Seal (1987).

Thornton, Ian W.B. 1978. "White Tiger Genetics—Further Evidence." *Journal of Zoology* (London), Vol. 185.

Thornton, Ian W.B., et al. 1967. "The Genetics of the White Tigers of Rewa." *Journal of Zoology* (London), Vol. 152.

Tilson, Ronald L., and Ulysses S. Seal, eds. 1987. *Tigers of the World.* Park Ridge, N.J.: Noyes Publications.

## TO LIVE AND DIE IN L.A.

APHIS. 1990. *Animal Damage Control Program.* Draft Environmental Impact Statement. Issued by the Animal and Plant Health Inspection Service, U.S. Department of Agriculture. Washington, D.C.

Bekoff, Marc, ed. 1978. *Coyotes: Biology, Behavior, and Management.* New York: Academic Press.

Crabtree, Robert L., and Ernest D. Ables. 1988. "Socio-Demographic Characteristics of an Unexploited Coyote Population in the Shrub-Steppe of Washington." Unpublished paper, University of Idaho, Department of Fish and Wildlife Resources.

Crabtree, Robert L., Jeffrey W. Blatt, and Kathleen A. Fulmer. 1988. "Social and Spatial Dynamics of an Unexploited Coyote Population in the Shrubsteppe of Washington." Unpublished paper, University of Idaho, Department of Fish and Wildlife Resources.

Gardetta, Dave. 1992. "The Permanent Coyote and the Disappearance of Wild L.A." *L.A. Weekly,* February 14–20, 1992.

Gier, H.T. 1975. "Ecology and Behavior of the Coyote (*Canis latrans*)." In

*The Wild Canids: Their Systematics, Behavioral Ecology and Evolution,* edited by M.W. Fox. New York: Van Nostrand Reinhold Company.

Gill, Don. 1970. "The Coyote and the Sequential Occupants of the Los Angeles Basin." *American Anthropologist,* Vol. 72.

Gill, Don, and Penelope Bonnett. 1973. *Nature in the Urban Landscape: A Study of City Ecosystems.* Baltimore: York Press.

Glustrom, Leslie. 1991. "Coyotes Shot from the Air Near Prescott, Arizona." Tucson, Ariz.: *Wildlife Damage Review,* No. 1, Summer 1991.

Halfpenny, James. 1986. *A Field Guide to Animal Tracking in Western America.* Boulder, Colo.: Johnson Books.

Milstein, Michael. 1991. "A Federal Killing Machine Rolls On." *High Country News,* January 28, 1991.

Schneider, Keith. 1991. "Mediating the Federal War on Wildlife." *New York Times,* June 9, 1991.

Silvestro, Roger L. di. 1985. "The Federal Animal Damage Control Program." In *Audubon Wildife Report 1985.* New York: The National Audubon Society.

Young, Stanley P., and Hartley H.T. Jackson. 1951. *The Clever Coyote.* Harrisburg, Pa.: The Stackpole Company.

## REACTION WOOD

Bell, Adrian D. 1991. *Plant Form: An Illustrated Guide to Flowering Plant Morphology.* Oxford: Oxford University Press.

Brockman, C. Frank. 1979. *Trees of North America.* New York: Golden Press.

Fisher, Jack B., and David E. Hibbs. 1982. "Plasticity of Tree Architecture: Specific and Ecological Variations Found in Aubreville's Model." *American Journal of Botany,* Vol. 69, No. 5.

Gartner, Barbara L., ed. 1995. *Plant Stems: Physiology and Functional Morphology.* San Diego: Academic Press.

Hallé, F., R.A.A. Oldeman, and P.B. Tomlinson. 1978. *Tropical Trees and Forests: An Architectural Analysis.* Berlin: Springer-Verlag.

Harlow, William M. 1970. *Inside Wood: Masterpiece of Nature.* Washington, D.C.: American Forestry Association.

Horn, Henry S. 1971. *The Adaptive Geometry of Trees.* Monographs in Population Biology, No. 3. Princeton, N.J.: Princeton University Press.

Mattheck, C. 1991. *Trees: The Mechanical Design.* Berlin: Springer-Verlag.

McMahon, Thomas. 1973. "Size and Shape in Biology." *Science,* Vol. 179, No. 4079. March 23, 1973.

McMahon, Thomas A. 1975. "The Mechanical Design of Trees." *Scientific American,* Vol. 233, No. 1, July 1975.

McMahon, Thomas A., and John Tyler Bonner. 1983. *On Size and Life.* New York: Scientific American Books.

McMahon, Thomas A., and Richard E. Kronauer. 1976. "Tree Structures: Deducing the Principle of Mechanical Design." *Journal of Theoretical Biology,* Vol. 59, No. 2.

Meylan, B.A., and B.G. Butterfield. 1972. *Three-Dimensional Structure of Wood: A Scanning Electron Microscope Study.* Syracuse, N.Y.: Syracuse University Press.

Sinnott, Edmund W. 1952. "Reaction Wood and the Regulation of Tree Form." *American Journal of Botany,* Vol. 39, No. 1.

Tomlinson, P.B. 1983. "Tree Architecture." *American Scientist,* Vol. 71, No. 2.

———. 1987 "Architecture of Tropical Plants." *Annual Review of Ecology and Systematics,* Vol. 18.

Walker, Aidan, ed. 1989. *The Encyclopedia of Wood.* Oxford: Facts on File.

Wilson, Brian F., and Robert R. Archer. 1977. "Reaction Wood: Induction and Mechanical Action." *Annual Review of Plant Physiology,* Vol. 28.

Zimmermann, Martin H., and Claud L. Brown. 1971. *Trees: Structure and Function.* New York: Springer-Verlag.

## SUPERDOVE ON 46TH STREET

Bodio, Stephen J. 1990. *Aloft: A Meditation on Pigeons and Pigeon-Flying.* New York: Lyons & Burford.

Coghlan, Andy. 1990. "Pigeons, Pests and People." *New Scientist,* December 1, 1990.

Darwin, Charles. 1859. *The Origin of Species.* Reprint edition: Avenel Books, New York, 1979.

———. 1868. *The Variation of Animals and Plants under Domestication.* Reprint edition: D. Appleton, New York, 1892.

Desmond, Adrian, and James Moore. 1991. *Darwin.* New York: Warner Books.

Feduccia, Alan. 1980. *The Age of Birds.* Cambridge, Mass.: Harvard University Press.

Goodwin, Derek. 1983. *Pigeons and Doves of the World.* Ithaca, N.Y.: Cornell University Press.

Johnston, Richard F. 1972. "Pigeons." In *Grzimek's Animal Life Encyclopedia,* Vol. 8.

———. 1990. "Variation in Size and Shape in Pigeons, *Columba livia.*" *The Wilson Bulletin,* Vol. 102, No. 2.

———. 1992. "Evolution in the Rock Dove: Skeletal Morphology." *Auk,* Vol. 109, No. 3.

———. 1994. "Geographic Variation of Size in Feral Pigeons." *Auk,* Vol. 111, No. 2.

Johnston, Richard F., Douglas Siegel-Causey, and Steven G. Johnson. 1988. "European Populations of the Rock Dove *Columba livia* and Genotypic Extinction." *American Midland Naturalist,* Vol. 120, No. 1.

Johnston, Richard F., and Marián Janiga. 1995. *Feral Pigeons.* New York: Oxford University Press.

Marshall, A.J., ed. 1960. *Biology and Comparative Physiology of Birds,* Vol. 1. New York: Academic Press.

Schorger, A.W. 1952. "Introduction of the Domestic Pigeon." *Auk,* Vol. 69.

Sossinka, Roland. 1982. "Domestication in Birds." In *Avian Biology,* Vol. 6. Edited by Donald S. Farner, James R. King, and Kenneth C. Parkes. New York: Academic Press.

Whitfield, Philip, ed. *The Macmillan Illustrated Encyclopedia of Birds.* New York: Macmillan.

Zeuner, Frederick E. 1963. *A History of Domesticated Animals.* New York: Harper & Row.

## BEFORE THE FALL

Barbosa, Pedro, and Jack C. Schultz, eds. 1987. *Insect Outbreaks.* New York: Academic Press, Inc.

Berryman, Alan A. 1987. "The Theory and Classification of Outbreaks." In Barbosa and Schultz (1987).

Catton, William R., Jr. 1982. *Overshoot: The Ecological Basis of Revolutionary Change.* Chicago: University of Illinois Press.

Clark, Edwin C. 1958. "Ecology of the Polyhedroses of Tent Caterpillars." *Ecology,* Vol. 39, No. 1, January 1958.

Colinvaux, Paul. 1980. *The Fates of Nations: A Biological Theory of History.* New York: Simon and Schuster.

Ehrlich, Paul R., and Anne H. Ehrlich. 1990. *The Population Explosion.* New York: Simon and Schuster.

Fitzgerald, T.D. 1976. "Trail Marking by Larvae of the Eastern Tent Caterpillar." *Science,* Vol. 194, November 26, 1976.

———. 1993. "Sociality in Caterpillars." In *Caterpillars: Ecological and Evolutionary Constraints on Foraging.* Edited by Nancy E. Stamp and Timothy M. Casey. New York: Chapman and Hall.

Fitzgerald, T.D., and James T. Costa III. 1986. "Trail-based Communication and Foraging Behavior of Young Colonies of Forest Tent Caterpillars (Lepidoptera: Lasiocampidae)." *Annals of the Entomological Society of America,* Vol. 79, No. 6, November 1986.

Fitzgerald, T.D., and S. C. Peterson. 1988. "Cooperative Foraging and Communication in Caterpillars." *BioScience,* Vol 38, No. 1. January 1988.

Greenblatt, J.A., and J.A. Witter. 1976. "Behavioral Studies on *Malacosoma disstria* (Lepidoptera: Lasiocampidae)." *The Canadian Entomologist,* Vol. 108, November 1976.

Hardin, Garrett, ed. 1964. *Population, Evolution, and Birth Control: A Collage of Controversial Readings.* San Francisco: W.H. Freeman and Company.

Hunter, Alison F. 1991. "Traits That Distinguish Outbreaking and Nonoutbreaking Macrolepidoptera Feeding on Northern Hardwood Trees." *Oikos,* Vol. 60, No. 3.

Huxley, Julian. 1963. *The Human Crisis.* Seattle: University of Washington Press.

Klein, David R. 1968. "The Introduction, Increase, and Crash of Reindeer on St. Matthew Island." *Journal of Wildlife Management,* Vol. 32, No. 2., April 1968.

Myers, Judith H. 1990. "Population Cycles of Western Tent Caterpillars: Experimental Introductions and Synchrony of Fluctuations." *Ecology,* Vol. 71, No. 3.

———. 1993. "Population Outbreaks in Forest Lepidoptera." *American Scientist,* Vol. 81, May–June 1993.

Nothnagle, Philip J., and Jack C. Schultz. 1987. "What Is a Forest Pest?" In Barbosa and Schultz (1987).

Rosenberg, Charles E. 1989. "What Is an Epidemic? AIDS in Historical Perspective." *Daedalus,* Vol. 118, No. 2.

Rosenzweig, Michael L. 1974. *And Replenish the Earth: The Evolution, Consequences, and Prevention of Overpopulation.* New York: Harper & Row.

Sippell, W.L. 1962. "Outbreaks of the Forest Tent Caterpillar, *Malacosoma disstria* Hbn., a Periodic Defoliator of Broad-leaved Trees in Ontario." *Canadian Entomologist,* Vol. 94, April 1962.

Stenseth, Nils C. 1987. "Evolutionary Processes and Insect Outbreaks." In Barbosa and Schultz (1987).

## PINHEAD SECRETS

Gillette, Ned, and John Dostal. 1988. *Cross-Country Skiing.* Seattle: The Mountaineers.

Goodman, David. 1989. *Classic Backcountry Skiing.* Boston: Appalachian Mountain Club.

Kleppen, Halvor. 1986. *Telemarkskiing: Norway's Gift to the World,* translated by Margaret Saltveit and Margaret Wold. Oslo: Det Norske Samlaget.

Parker, Paul. 1988. *Free-Heel Skiing.* Chelsea, Vt.: Chelsea Green Publishing Co.

## THE KEYS TO KINGDOM COME

"Space and Missile Orientation Course Study Guide." Vandenberg Air Force Base, Calif.: U.S. Air Force, n.d. Private circulation, to say the least.

Nozick, Robert. 1985. "Newcomb's Problem and Two Principles of

Choice." In *Paradoxes of Rationality and Cooperation: Prisoner's Dilemma and Newcomb's Problem,* edited by Richmond Campbell and Lanning Sowden. Vancouver: The University of British Columbia Press.

## KARL'S SENSE OF SNOW

Atwater, Montgomery M. 1968. *The Avalanche Hunters.* Philadelphia: Macrae Smith Company.

Bentley, W.A., and W.J. Humphreys. 1962. *Snow Crystals.* New York: Dover Publications, Inc.

Birkeland, K.W., K.J. Hansen, and R.L. Brown. 1995. "The Spatial Variability of Snow Resistance on Potential Avalanche Slopes." *Journal of Glaciology,* Vol. 41, No. 137.

Burtscher, Martin. 1994. "Avalanche Survival Chances." *Nature,* Vol. 371, October 6, 1994.

Casimiro, Steve. 1994. "State of the Art: Avalanche Control." *Powder,* December 1994.

Falk, Markus, Hermann Brugger, and Liselotte Adler-Kastner. 1994. "Avalanche Survival Chances." *Nature,* Vol. 368. March 3, 1994.

Fraser, Colin. 1966. *The Avalanche Enigma.* Chicago: Rand McNally & Co.

Fredston, Jill A., and Doug Fesler. 1994. *Snow Sense: A Guide to Evaluating Snow Avalanche Hazard.* Anchorage: Alaska Mountain Safety Center, Inc.

Høeg, Peter. 1993. *Smilla's Sense of Snow,* translated by Tiina Nunnally. New York: Bantam Doubleday Dell.

Jamieson, J.B., and C.D. Johnston. 1993. "Rutschblock Precision, Technique Variations and Limitations." *Journal of Glaciology,* Vol. 39, No. 133.

Johnson, Ron, and Karl Birkeland. 1994. "The Stuffblock: A Simple and Effective Snowpack Stability Test." Paper presented at the 1994 International Snow Science Workshop, Snowbird, Utah. Ms. from the authors.

Kirk, Ruth. *Snow.* 1978. New York: William Morrow.

LaChapelle, Edward R. 1980. "The Fundamental Processes in Conventional Avalanche Forecasting." *Journal of Glaciology,* Vol. 26, No. 94.

McClung, David, and Peter Schaerer. 1993. *The Avalanche Handbook.* Seattle: The Mountaineers.

Montagne, John. 1980. "The University Course in Snow Dynamics—A Stepping-Stone to Career Interests in Avalanche Hazards." *Journal of Glaciology,* Vol. 26, No. 94.

Perla, Ronald I., and M. Martinelli, Jr. 1978. *Avalanche Handbook.* Agriculture Handbook 489. Fort Collins, Colo.: U. S. Department of Agriculture/Forest Service.

Seligman, G. 1936. *Snow Structure and Ski Fields.* London: Macmillan.

Williams, Knox, and Betsy Armstrong. N.d. *The Snowy Torrents: Avalanche Accidents in the United States 1972–79.* Jackson, Wyo.: Teton Bookshop Publishing Co.

Wilson, Norman A. 1977. "Everything You Wanted to Know About Snow." *Mariah,* Winter 1977.

## THE TREES CRY OUT ON CURRAWONG MOOR

Clark, Julia. 1983. *The Aboriginal People of Tasmania.* Hobart: Tasmanian Museum and Art Gallery.

Kirkpatrick, J.B., and Sue Backhouse. 1985. *Native Trees of Tasmania.* Hobart: Sue Backhouse.

Lennox, Geoff. N.d. "The Van Diemens Land Company and the Tasmanian Aborigines: A Reappraisal." Uncorrected draft ("basically accurate but uncorrected," Lennox calls it, and asks that it be cited as "discarded draft") of a paper, written by Lennox as staff historian with the Tasmanian Department of Parks, Wildlife and Heritage, Hobart.

Plomley, N.J.B. 1977. *The Tasmania Aborigines.* Launceston: published by the author.

———. 1987. *Weep in Silence: A History of the Flinders Island Aboriginal Settlement.* Hobart: Blubber Head Press.

———, ed. 1966. *Friendly Mission: The Tasmanian Journals and Papers of George Augustus Robinson, 1829–1834.* Hobart: Tasmanian Historical Research Association.

Robson, Lloyd. 1985. *A Short History of Tasmania.* Melbourne: Oxford University Press.

Ryan, Lyndall. 1981. *The Aboriginal Tasmanians.* St. Lucia: University of Queensland Press.

Turnbull, Clive. 1948. *Black War: The Extermination of the Tasmanian Aborigines.* Melbourne: F.W. Cheshire.

## THE BIG TURN

Dostal, John. "Dick Hall on Backcountry Skiing: Fooling the Fall Line." *Back Country.* December 1994/January 1995.

Telemark Movies, Inc. *The Telemark Movie, Revenge of the Telemarkers,* and *Telemark Workshop.* Videos featuring the technical teachings and *esprit de ski* of Dickie Hall. Waitsfield, Vt.: North American Telemark Organization.

## EAT OF THIS FLESH

Braun, Clait E., ed. 1991. *Mountain Lion–Human Interactions.* Proceedings of a symposium and workshop, April 24–26, 1991. Published by the Colorado Division of Wildlife.

Hansen, Kevin. 1992. *Cougar.* Flagstaff, Ariz.: Northland Publishing Company.

Kerasote, Ted. 1993. *Bloodties.* New York: Random House.

Linn, Amy. 1993. "Wild Cats Wild." *Audubon,* July/August 1993.

Lopez, Barry. 1981. "An Elusive Cat." *GEO,* June 1981.

Nelson, Richard. 1993. "Searching for the Lost Arrow: Physical and Spiritual Ecology in the Hunter's World." In *The Biophilia Hypothesis,* edited by Stephen R. Kellert and Edward O. Wilson. Washington, D.C.: Island Press/Shearwater Books.

Riley, Shawn J. 1992. "Cougars in Montana: A Review of Biology and Management and a Plan for the Future—Draft." Helena, Mont.: Montana Department of Fish, Wildlife and Parks.

Shaw, Harley. 1989. *Soul Among Lions.* Boulder, Colo.: Johnson Books.

Thomas, E. Donnall, Jr. 1993. "To Skin a Cat." *Bowhunter,* February/March 1993.

———. 1994. "The Idea of Order in Cougar Country." *Traditional Bowhunter,* December/January 1994.

———. 1993. *Longbows in the Far North.* Mechanicsburg, Pa.: Stackpole Books.

## THE SWALLOW THAT HIBERNATES UNDERWATER

Alerstam, Thomas. 1990. *Bird Migration,* translated by David A. Christie. Cambridge: Cambridge University Press.

Aristotle. 1956. *Historia Animalium.* Vol. 4 of *The Works of Aristotle,* translated by D'Arcy Wentworth Thompson, edited by J.A. Smith and W.D. Ross. Oxford: The Clarendon Press.

Browne, Janet. 1983. *The Secular Ark: Studies in the History of Biogeography.* New Haven, Conn.: Yale University Press.

Dorst, Jean. 1974. *The Life of Birds,* Vol. 2, translated by I.C.J. Galbraith. New York: Columbia University Press.

Gwinner, E., ed. 1990. *Bird Migration: Physiology and Ecophysiology.* Berlin: Springer-Verlag.

Jenkins, Alan C. 1978. *The Naturalists: Pioneers of Natural History.* New York: Mayflower Books.

Mabey, Richard. 1986. *Gilbert White: A Biography of the Author of "The Natural History of Selborne."* London: Century Hutchinson Ltd.

Mead, Chris. 1983. *Bird Migration.* New York: Facts on File Publications.

Moorhead, Alan. 1987. *The Fatal Impact: The Invasion of the South Pacific 1767–1840.* Sydney: Mead & Beckett Publishing.

White, Gilbert. 1977. *The Natural History of Selborne,* edited with an introduction and notes by Richard Mabey. Harmondsworth, Middlesex, Eng.: Penguin Books. Originally published in 1788.

———. 1985. *The Essential Gilbert White of Selborne.* Edited by H.J. Massingham; selected and introduced by Mark Daniel. Boston: David R. Godine.

Worster, Donald. 1985. *Nature's Economy: A History of Ecological Ideas.* Cambridge: Cambridge University Press.

## TRINKET FROM ARU

Beehler, Bruce M. 1989. "The Birds of Paradise." *Scientific American,* Vol. 261, December 1989.

Beehler, Bruce M., Thane K. Pratt, and Dale A. Zimmerman. 1986. *Birds of New Guinea.* Princeton, N.J.: Princeton University Press.

Diamond, Jared M. 1981. "Birds of Paradise and the Theory of Sexual Selection." *Nature,* Vol. 293, September 24, 1981.

Dinsmore, James J. 1970. "Courtship Behavior of the Greater Bird of Paradise." *Auk,* Vol. 87, April 1970.

LeCroy, Mary. 1981. "The Genus *Paradisaea*—Display and Evolution." *American Museum Novitates,* No. 2714, June 24, 1981.

Wallace, Alfred Russel. 1857. "On the Natural History of the Aru Islands." *The Annals and Magazine of Natural History,* December 1857.

———. 1962. *The Malay Archipelago.* New York: Dover Publications, Inc. Originally published in London: Macmillan, 1869.

## BAGPIPES FOR ED

Abbey, Edward. 1971. *Desert Solitaire: A Season in the Wilderness.* New York: Ballantine Books. Originally published by Simon and Schuster, 1968.

## POINT OF ATTACHMENT

Barnes, H., and Margaret Barnes. 1957. "Resistance to Desiccation in Intertidal Barnacles." *Science,* Vol. 126, August 23, 1957.

Barnes, H., and E.S. Reese. 1960. "The Behaviour of the Stalked Intertidal Barnacle *Pollicipes polymerus* J.B. Sowerby, with Special Reference to Its Ecology and Distribution." *Journal of Animal Ecology,* Vol. 29.

Brent, Peter. 1981. *Charles Darwin: A Man of Enlarged Curiosity.* New York: W.W. Norton.

Brusca, Richard C., et al. 1980. *Common Intertidal Invertebrates of the Gulf of California.* Tucson: The University of Arizona Press.

Connell, Joseph H. 1961. "The Influence of Interspecific Competition and Other Factors on the Distribution of the Barnacle *Chthamalus stellatus.*" *Ecology,* Vol. 42, No. 4.

———. 1970. "A Predator-Prey System in the Marine Intertidal Region. I. *Balanus glandula* and Several Predatory Species of *Thais.*" *Ecological Monographs,* Vol. 40, No. 1.

———. 1972. "Community Interactions on Marine Rocky Intertidal Shores." *Annual Review of Ecology and Systematics,* Vol. 3.

Darwin, Charles. 1851. *A Monograph on the Sub-Class Cirripedia.* Vol. 1: "The Lepadidae; or, Pedunculated Cirripedes." London: The Ray Society.

———. 1854. *A Monograph on the Sub-Class Cirripedia.* Vol. 2: "The Balanidae, (or, Sessile Cirripedes); The Verrucidae, Etc. Etc. Etc." London: The Ray Society.

———. 1969. *The Autobiography of Charles Darwin, 1809–1882,* edited by Nora Barlow. New York: W.W. Norton.

Dayton, Paul K. 1971. "Competition, Disturbance, and Community Organization: The Provision and Subsequent Utilization of Space in a Rocky Intertidal Community." *Ecological Monographs,* Vol. 41, No. 4.

Desmond, Adrian, and James Moore. 1991. *Darwin.* New York: Warner Books.

Ghiselin, Michael T. 1984. *The Triumph of the Darwinian Method.* Chicago: The University of Chicago Press.

Gould, Stephen Jay. 1977. "Darwin's Delay," in *Ever Since Darwin.* New York: W.W. Norton.

Hilgard, Galen Howard. 1960. "A Study of Reproduction in the Intertidal Barnacle, *Mitella polymerus,* in Monterey Bay, California." *The Biological Bulletin,* Vol. 119, No. 2.

Himmelfarb, Gertrude. 1968. *Darwin and the Darwinian Revolution.* New York: W.W. Norton.

Howard, Galen Kent, and Henry C. Scott. 1959. "Predaceous Feeding in Two Common Gooseneck Barnacles." *Science,* Vol. 129, March 13, 1959.

Johnson, Myrtle Elizabeth, and Harry James Snook. 1927. *Seashore Animals of the Pacific Coast.* New York: Macmillan.

Knight-Jones, E.W. 1955. "The Gregarious Setting Reaction of Barnacles as a Measure of Systematic Affinity." *Nature,* Vol. 175, No. 4449, February 5, 1955.

Knight-Jones, E.W., and D.J. Crisp. 1953. "Gregariousness in Barnacles in Relation to the Fouling of Ships and to Anti-Fouling Research." *Nature,* Vol. 171, No. 4364, June 20, 1953.

Kozloff, Eugene N. 1993. *Seashore Life of the Northern Pacific Coast.* Seattle: University of Washington Press.

Lewis, Cindy Arey, and Fu-Shiang Chia. 1981. "Growth, Fecundity, and Reproductive Biology in the Pedunculate Cirripede *Pollicipes polymerus* at San Juan Island, Washington." *Canadian Journal of Zoology,* Vol. 59.

Newell, R.C. 1970. *Biology of Intertidal Animals.* New York: American Elsevier Publishing Company.

Ricketts, Edward F., Jack Calvin, and Joel W. Hedgpeth; revised by David W. Phillips. 1985. *Between Pacific Tides.* Stanford, Calif.: Stanford University Press.

## VOICE PART FOR A DUET

Daly, Martin, and Margo Wilson. 1983. *Sex, Evolution, and Behavior.* Boston: Willard Grant Press.

Dunbar, Robin. 1984. "The Ecology of Monogamy." *New Scientist,* Vol. 103, August 30, 1984.

———. 1985. "Monogamy on the Rocks." *Natural History,* Vol. 94, No. 11, November 1985.

Dunbar, R.I.M., and E.P. Dunbar. 1980. "The Pairbond in Klipspringer." *Animal Behaviour,* Vol. 28.

Fisher, Helen E. 1992. *Anatomy of Love: The Natural History of Monogamy, Adultery, and Divorce.* New York: W.W. Norton.

Hrdy, Sarah Blaffer. 1981. *The Woman That Never Evolved.* Cambridge, Mass.: Harvard University Press.

Kleiman, Devra G. 1977. "Monogamy in Mammals." *The Quarterly Review of Biology,* Vol. 52, March 1977.

Lack, David. 1968. *Ecological Adaptations for Breeding in Birds.* London: Methuen & Co.

Mason, William A. 1974. "Comparative Studies of Social Behavior in *Callicebus* and *Saimiri:* Behavior of Male-Female Pairs." *Folia Primatologica,* Vol. 22.

Müller-Schwarze, Dietland. 1984. *The Behavior of Penguins: Adapted to Ice and Tropics.* Albany: State University of New York Press.

Orians, Gordon H. 1969. "On the Evolution of Mating Systems in Birds and Mammals." *American Naturalist,* Vol. 103, No. 934.

Pollock, J.I. 1975. "Field Observations on *Indri indri:* A Preliminary Report." In *Lemur Biology,* Ian Tattersall and Robert W. Sussman, eds. New York: Plenum Press.

Pooley, Anthony C., and Carl Gans. 1976. "The Nile Crocodile." *Scientific American,* Vol. 234, No. 4, April 1976.

Rutberg, Allen T. 1983. "The Evolution of Monogamy in Primates." *Journal of Theoretical Biology*, Vol. 104.

Smith, J. Maynard. 1977. "Parental Investment: A Prospective Analysis." *Animal Behaviour*, Vol. 25.

Tilson, Ronald L., and Richard R. Tenaza. 1976. "Monogamy and Duetting in an Old World Monkey." *Nature*, Vol. 263, September 23, 1976.

Tilson, Ronald L., and Peter M. Norton. 1981. "Alarm Duetting and Pursuit Deterrence in an African Antelope." *American Naturalist*, Vol. 118, No. 3.

Van Schaik, C.P., and R.I.M. Dunbar. 1990. "The Evolution of Monogamy in Large Primates: A New Hypothesis and Some Crucial Tests." *Behavior*, Vol. 115, Nos. 1–2.

Verner, Jared. 1964. "Evolution of Polygamy in the Long-Billed Marsh Wren." *Evolution*, Vol. 18, No. 2.

Verner, Jared, and Mary F. Willson. 1966. "The Influence of Habitats on Mating Systems of North American Passerine Birds." *Ecology*, Vol. 47, No. 1.

Walters, Mark Jerome. 1989. *Courtship in the Animal Kingdom*. New York: Anchor Books.

Wilson, Edward O. 1980. *Sociobiology*, abridged edition. Cambridge, Mass.: Belknap Press of Harvard University Press.

Wittenberger, James F., and Ronald L. Tilson. 1980. "The Evolution of Monogamy: Hypotheses and Evidence." *Annual Review of Ecology and Systematics*, Vol. 2.

## LOVE IN THE AGE OF RELATIVITY

Asimov, Isaac. 1971. *The Universe: From Flat Earth to Quasar*. Harmondsworth, Middlesex, Eng.: Penguin Books.

Derry, T. K., and Trevor I. Williams. 1960. *A Short History of Technology*. London: Oxford University Press.

Einstein, Albert. 1961. *Relativity: The Special and the General Theory*, translated by Robert W. Lawson. New York: Crown Publishers.

Einstein, Albert, and Leopold Infeld. 1938. *The Evolution of Physics: From Early Concepts to Relativity and Quanta*. New York: Simon and Schuster.

Fraser, J.T. 1982. *The Genesis and Evolution of Time: A Critique of Interpretation in Physics*. Amherst: University of Massachusetts Press.

Gould, Stephen Jay. 1987. *Time's Arrow Time's Cycle: Myth and Metaphor in the Discovery of Geological Time*. Cambridge, Mass.: Harvard University Press.

Hawking, Stephen W. 1988. *A Brief History of Time: From the Big Bang to Black Holes*. New York: Bantam Books.

Kaufmann, William J., III. 1977. *Relativity and Cosmology*. New York: Harper and Row.

Landes, David S. 1983. *Revolution in Time: Clocks and the Making of the*

*Modern World.* Cambridge, Mass.: Belknap Press of Harvard University Press.

Lopez, Barry. 1981. *Winter Count.* New York: Charles Scribner's Sons.

Macey, Samuel L. 1980. *Clocks and the Cosmos: Time in Western Life and Thought.* Hamden, Conn.: Archon Books.

Mumford, Lewis. 1963. *Technics and Civilization.* New York: Harcourt, Brace & World.

Price, Derek de Solla. 1975. *Science Since Babylon.* New Haven, Conn.: Yale University Press.

Russell, Bertrand. 1959. *The ABC of Relativity.* New York: New American Library.

Shallis, Michael. 1983. *On Time: An Investigation into Scientific Knowledge and Human Experience.* New York: Schocken Books.

Withrow, G.J. 1973. *The Nature of Time.* New York: Holt, Rinehart and Winston.

Wright, Lawrence. 1968. *Clockwork Man.* London: Elek Books Ltd.

## STRAWBERRIES UNDER ICE

Childress, Donald Arthur. 1971. "Canada Goose Production and Water Level Relationships on the Madison River, Montana." Bozeman, Mont.: Master's thesis, Montana State University.

Dyson, James L. 1962. *The World of Ice.* New York: Alfred A. Knopf.

Leontiev, Konstantin. 1969. *Against the Current: Selections from the Novels, Essays, Notes, and Letters of Konstantin Leontiev.* Edited by George Ivask, translated from the Russian by George Reavey. New York: Weybright and Talley.

Lopez, Barry. 1986. *Arctic Dreams: Imagination and Desire in a Northern Landscape.* New York: Charles Scribner's Sons.

Matthews, Samuel W. 1987. "Ice on the World." *National Geographic,* Vol. 171, No. 1, January 1987.

Parfit, Michael. 1985. *South Light: A Journey to the Last Continent.* New York: Macmillan.

Pyne, Stephen J. 1986. *The Ice: A Journey to Antarctica.* Iowa City: University of Iowa Press.

Spray, James S. 1956. "Early Days in the Madison Valley: A History." Unpublished typescript in the library of Montana State University, Bozeman. Compiled anonymously from an earlier (ca. 1936) manuscript by the author.

Warren, Robert Penn. 1969. *Audubon: A Vision.* New York: Random House.

# Index